T0419456

SPACE SCIENCE, EXPLORATION AND POLICIES

SPACE POLICY AND ITS RAMIFICATIONS

SPACE SCIENCE, EXPLORATION AND POLICIES

Additional books in this series can be found on Nova's website under the Series tab.

Additional E-books in this series can be found on Nova's website under the E-books tab.

SPACE SCIENCE, EXPLORATION AND POLICIES

SPACE POLICY AND ITS RAMIFICATIONS

JOHN P. RAMOS
EDITOR

Nova Science Publishers, Inc.
New York

Additional color graphics may be available in the e-book version of this book.

LIBRARY OF CONGRESS CATALOGING-IN-PUBLICATION DATA
Space policy and its ramifications / editor, John P. Ramos.
p. cm.
Includes index.
ISBN 978-1-61761-555-9 (hardcover)
1. Astronautics--Government policy--United States. 2.
Astronautics--Government policy--Canada. 3. Astronautics--Government
policy--Korea. 4. Astronautics and state. I. Ramos, John P.
TL789.8.U5S5875 2010
629.4--dc22
2010031348

Published by Nova Science Publishers, Inc. † New York

CONTENTS

PREFACE

For the past several years, the priorities of NASA have been governed by the Vision for Space Exploration. The Vision was announced by President Bush in January 2004 and endorsed by Congress in the 2005 and 2008 NASA authorization acts. It directed NASA to focus its efforts on returning humans to the Moon by 2020 and some day sending them to Mars and "worlds beyond." The resulting efforts are now approaching major milestones, such as the end of the space shuttle program, design review decisions for the new spacecraft intended to replace the shuttle, and decisions about whether to extend the operation of the International Space Station. At the same time, concerns have grown about whether NASA can accomplish the planned program of human exploration of space without significant growth in its budget. This book explores the current U.S. space policy and its ramifications.

Chapter 1 - Safe pipeline laying-out and transportation of energy resources are a major concern for the public and the pipeline industry.

There are several aspects for support of critical infrastructure protection using the remote sensing for oil and gas pipeline safety purposes. It is an improvement of the transportation security, social and economical benefits, natural disaster, risk to property and human life etc.

Different countries use of different regulations and instructions for management of actions and the monitor of pipelines for transportation of natural oil and gas. Irrespective of details of individual requirements, it is an interest of any owner of the pipeline to support the pipeline in working condition and protect them effectively against the damage caused by the third parties. The general methods of monitoring most widely used for pipelines of transportation of natural oil and gas include patrolling among the route of the pipeline and remote observation from the appropriate height. For realization of monitoring of pipelines it is widely used small planes and helicopters. Though these methods are the high guarantor of safety of oil and gas pipelines but in this case the price is also enough high. Significant progress in the field of technology of high resolution remote sensing data used as plane and satellite onboard sensors and reduction of the cost due to increase of a memory capacity of computer technology and expansion of an opportunity of data processing technology was created with favorable conditions of application of this technology in safety and security issues of pipelines in transportations of oil and gas.

Today opportunity of high resolution of space imagery use creates a positive environment on application of space technology for monitoring of integrated systems in different areas of industry and commercial purposes. They are an airborne, unmanned aerial vehicles (UAV) and satellite Synthetic Aperture Radar and LIDAR multi-spectral and hyper-spectral sensors.

One of the advance methods of feasibility study of appropriate services for monitoring of integration systems is based on remote sensing data and Geographic Information System (GIS) developments. Objective of this approach is to improve safety, security aspects of integration systems, reduce survey costs and improve transportation and transmission efficiency through an increased monitoring frequency.

Earth observation for investigation requires very high resolution optical and in most cases radar sensors. A very high resolution is here defined as a resolution of 1 meter and better. This scale is required to allow the detection of targets. A very high resolution satellite observation, formerly limited to national and strictly classified reconnaissance tasks, has become a commercial business in the recent years. Affected by satellites losses and pioneered with Space Imaging's IKONOS satellite, better than 1 meter resolution optical data is now available for science and commercial use.

The use of advance technology makes available to enhance planning, design, management, operation and maintenance of the linear infrastructure. Aerial and satellite data based on the remote sensing methods with further integration to GIS technologies represent the area of rapid responds and developments which can be leveraged to assist the linear systems risk assessment to assure the safety of linear infrastructures. Industrial and scientific achievements in satellite remote sensing methods and appropriate data processing techniques are opening a new wide technological opportunities and availabilities to develop an enhanced capability to accomplish the linear systems mapping and safety needs of the oil and gas transportation. Combination of those technologies with GIS has a valuable and significant potential for further application to a huge of cross cutting system of security issues.

Chapter 2 - For the past several years, the priorities of the National Aeronautics and Space Administration (NASA) have been governed by the Vision for Space Exploration. The Vision was announced by President Bush in January 2004 and endorsed by Congress in the 2005 and 2008 NASA authorization acts (P.L. 109-155 and P.L. 110-422). It directed NASA to focus its efforts on returning humans to the Moon by 2020 and some day sending them to Mars and "worlds beyond." The resulting efforts are now approaching major milestones, such as the end of the space shuttle program, design review decisions for the new spacecraft intended to replace the shuttle, and decisions about whether to extend the operation of the International Space Station. At the same time, concerns have grown about whether NASA can accomplish the planned program of human exploration of space without significant growth in its budget.

A high-level independent review of the future of human space flight, chaired by Norman R. Augustine, issued its final report in October 2009. It presented several options as alternatives to the Vision and concluded that for human exploration to continue "in any meaningful way," NASA would require an additional $3 billion per year above current plans.

In its FY2011 budget request, the Obama Administration proposed cancelling the Constellation spacecraft development program and eliminating the goal of returning humans to the Moon. NASA would instead rely on commercial providers to transport astronauts to Earth orbit, and its ultimate goal beyond Earth orbit would be human exploration of Mars, with missions to other destinations, such as visiting an asteroid in 2025, as intermediate goals. Operation of the International Space Station would be extended to at least 2020, and long-term technology development would receive increased emphasis.

Committees in the House and Senate have held hearings to consider both the Augustine report and the Administration proposals. As Congress considers these broad space policy challenges, it faces choices about whether NASA's human exploration program is affordable and sufficiently safe,and if so, what destination or destinations it should explore; whether the space shuttle program should continue past its currently plannedtermination at the end of 2010 (or in early 2011); if so, how to ensure thecontinued safety of shuttle crews after 2010; if not, how the transition of theshuttle workforce and facilities should be managed; whether U.S. use of the International Space Station should continue past itscurrently planned termination at the end of 2015; whether the currently planned Orion crew capsule and Ares rockets, beingdeveloped as successors to the space shuttle, are the best choices for deliveringastronauts and cargo into space, or whether other proposed rockets or commercialservices should take their place; and how NASA's multiple objectives in human spaceflight, science, aeronautics, and education should be prioritized.

Chapter 3 - This chapter analyzes the feasibility of research and development related to select space security technology development plans. The proposals are presented in an international context where certain interest groups in the U.S.A. are not likely to be alone in having aspirations to space security technology. China, Russia, and France, among others, are just as likely to develop novel space security technology. Since the U.S. programs have received more attention and were arguably in more advanced stages of planning, if not development, this chapter focuses on U.S. national security space management and organization proposals, investigating five key technology projects that have been proposed and partially pursued. These systems are described in relation to the technical issues, engineering difficulties and physical limits involved. If appropriate, their implications are linked to the political impact of implementing these plans to technological maturity. Various different countries could conceivably answer the perceived threat inherent in the development of space security technology by emulating or outpacing each other in space militarization. While the U.S.A. arguably has more advanced security technology expertise, it is shown that other countries have already deployed some security technologies in space applications in the past, with less media attention. And, ultimately, the nations with the most space assets also have the most to lose from space militarization.

Chapter 4 - It seems almost shocking to those who remember first hand the launch of Sputnik in October 1957 that the first fifty years of human space activities are now complete and then some. Certainly, more than a half-century of space-related experience represents enough time for an honest and thoroughgoing critical analysis of what we have now accomplished in space exploration, applications and science. Perhaps it is even more important to take this longer-term perspective to assess honestly where, why and how we have failed. This is a critical step toward assessing what future goals should be set for space in the coming half century. Such an exercise should let us consider who should be in charge of setting our future objectives in terms of space policy and a viable regulatory framework going forward. This should particularly focus on whether international, regional or national actors will be in charge of implementing and enforcing these space-related policies and regulations. It should also address what appears to be divergent-- as opposed to convergent— space policy perspectives as held by the various civil space agencies, the military or defense-related agencies, and the growing number of commercial entities? These three groups of critical actors in the space arena seem, in many instances, to have quite different views on how to best proceed both in terms of implementing space programs and activities and what are the best forms of space policy and regulation.

Key questions include the following: Should quite expensive human space exploration be our primary goal in either a global or national context? Should practical space applications—related to communications, remote sensing, climate change, meteorology, space navigation, disaster assessment and recovery and solar power-- represent a priority goal over human space exploration? Should deep space missions be essentially restricted to robotic systems and sensors to carry out space science? Should commercial space ventures be encouraged to thrive and supplant civil governmental space activities wherever and whenever possible—or at least in near Earth orbit? What about the strategic uses of space for military and defense related purposes? Should there be limits to the militarization of space and what form of collaboration, cooperation or formal divisions should there be with regard to commercial, civil governmental and military space operations? Who should be in charge of space safety regulations and their implementation?

These and a wide range of other questions seem ripe for serious analysis, study and future goal- setting purposes. In many ways these questions, for the next half century, may well come down to who leads and who follows? Will the leaders in space activities and space policy formation be the "Merchants" (the commercial space entrepreneurs), the "Guardians" (those that utilize space systems for military and defense-related purposes), or the "Civil Space Advocates & Regulators" (the official national and regional space agencies and regulatory authorities)? Which of these will become the dominant force in space—especially from Clarke Orbit (geosynchronous orbit) on down? Which of these "actors", working in tandem (or not) with the United Nations and other international agencies, will ultimately decide how we prioritize our outer space activities. In short, going forward, who will set space policies and regulations and who will "enforce" them? This, of course, is a lot for one article to try to address.

Thus the purpose of this article is to focus on one particular aspect of these longer-term outer space-related issues. This is to analyze the often different perspectives of the so-called "merchants", "guardians" and "civil space advocates and regulators" when it comes to space policy and regulation in some specific programmatic areas.

In broad brush perspective, many of the "merchants" as reflected by such groups as the "Commercial Spaceflight Federation" is for an open and "laissez faire" approach to the regulation of outer space in and around Earth. This perspective is in some ways like the "Law of the Seas" approach which suggests that once one is beyond national airspace limits (i.e. the equivalent to and above national territorial waters adjacent to a nation's landmass) there is a minimum of regulation beyond basic "rules of the road". This "open approach" would have every national or commercial entity taking care of their own projects' safety. The "guardians" have accepted the current broad and vague U.N. Outer Space Treaty (formally known as the *Treaty on Principles Governing the Activities of States in the Exploration and Use of Outer Space, including the Moon and Other Celestial Bodies)* This treaty indicates that there should not be "militarization of space", but leaves the precise definition of what militarization means wide open. Article IV does not explicitly ban "manned military spacecraft", although it does specify that no nuclear weapons are to be launched into orbit. Certainly within the current wording of the Outer Space Treaty military entities have deployed dedicated and dual-use communication satellites for military purposes as well as navigational satellites for missile targeting and Earth observation satellites for surveillance and intelligence gathering systems. Sub-orbital regional and intercontinental missile systems carrying weapons of mass destruction still exist and planning and engineering for so-called "star wars-like" systems for defense and "potentially offensive" purposes have been undertaken. When the UN General Assembly took a vote in 2008 on a measure entitled "Prevention of An Arms Race in Outer Space," 166 nations voted in favor of this resolution with two abstentions and the U.S. being just the one lone nation voting against the resolution.

Finally there are the civil space agencies of the world. These agencies are in synch with national governments and international agencies such as the United Nations Committee on the Peaceful Uses of Outer Space, the International Telecommunication Union, etc. The major space agencies have generally worked toward an internationally harmonized approach to outer space. This has been seen in close cooperation around the International Space Station (ISS), the new "Global Exploration Strategy" (in progress since 2004) that spells out the rationales for carrying out a long-term program of human space exploration, and cooperation within the United Nations to develop voluntary principles to limit the growth of orbital space debris.

In short, many of the "space merchants" would like to see a wide-open, entrepreneurial approach to space and space regulation. The "space guardians" would like to see a minimum of new military or defense-related restrictions on what can be done in outer space and yet have serious concerns about what private space entrepreneurs might be allowed to do in terms of operating private space planes or private space stations. The "civil space agencies" generally see the need for more international cooperation in space, common rules, regulations and standards for space safety, and an improved regulatory framework for cooperative space ventures.

The analysis that follows examines several emerging future space-related programs and activities. The focus of the analysis is how the "merchants", "the guardians" and "civil space advocates" might wish these activities to be regulated in terms of strict controls or much looser and free-form guidelines.

Chapter 5 - Korea's space development program was created almost 40 years behind advanced countries Nevertheless; it has been making remarkable growth including the construction of Korea 'NARO' space center and the possession of 10 satellites developed by Korea's own technology.

There are; however, a number of problems in Korean space development policy—dependence on imported core technology, uncertainty of budgetary allocations that are influenced by political or economical circumstances, insufficiency of human resources, concentrated authority on KARI(Korea Aerospace Research Institute), and weakness of aerospace industry infrastructures. They all must be re-aligned in order to develop long-term space industry strategies.

In this study, the authors analyzed the current space policy issues mentioned above by comparing Korea and other countries' space industries and government policies. In addition, we have suggested that the private sector should vitalize the space industry in terms of establishing a functional space industrial cluster. That is, Korea should establish a functional space industry cluster and remove inefficient, overlapping investments.

For successful space development, late comer countries should institute a privately financed space development policy. The establishment of a space industrial cluster policy in this study is expected to help invigorate the space industry of Korea and catch-up countries with a similar environment.

Chapter 6 - The "space age" began on October 4, 1957, when the Soviet Union (USSR) launched Sputnik, the world's first artificial satellite. Some U.S. policymakers, concerned about the USSR's ability to launch a satellite, thought Sputnik might be an indication that the United States was trailing behind the USSR in science and technology. The Cold War also led some U.S. policymakers to perceive the Sputnik launch as a possible precursor to nuclear attack. In response to this "Sputnik moment," the U.S. government undertook several policy actions, including the establishment of the National Aeronautics and Space Administration (NASA) and the Defense Advanced Research Projects Agency (DARPA), enhancement of research funding, and reformation of science, technology, engineering and mathematics (STEM) education policy.

Following the "Sputnik moment," a set of fundamental factors gave "importance, urgency, and inevitability to the advancement of space technology," according to an Eisenhower presidential committee. These four factors include the compelling need to explore and discover; national defense; prestige and confidence in the U.S. scientific, technological, industrial, and military systems; and scientific observation and experimentation to add to our knowledge and understanding of the Earth, solar system, and universe. They are still part of current policy discussions and influence the nation's civilian space policy priorities — both in terms of what actions NASA is authorized to undertake and the appropriations each activity within NASA receives. NASA has active programs that address all four factors, but many believe that it is being asked to accomplish too much for the available resources.

Further, the United States faces a far different world today. No Sputnik moment, Cold War, or space race exists to help policymakers clarify the goals of the nation's civilian space program. The Hubble telescope, Challenger and Columbia space shuttle disasters, and Mars exploration rovers frame the experience of current generations, in contrast to the Sputnik launch and the U.S. Moon landings that form the experience of older generations. As a result, some experts have called for new 21st century space policy objectives and priorities to replace those developed 50 years ago.

The authorization of NASA funding in the National Aeronautics and Space Act of 2005 (P.L. 109-55) extends through FY2008. Congress may decide to maintain or shift NASA's priorities during the next reauthorization. For example, if Congress believes that national prestige should be the highest priority, they may choose to emphasize NASA's human exploration activities, such as establishing a Moon base and landing a human on Mars. If they consider scientific knowledge the highest priority, unmanned missions and other science-related activities may be Congress' major goal for NASA. If international relations are a high priority, Congress might encourage other nations to become equal partners in NASA's activities. If spinoff effects, such as the creation of new jobs and markets and its effect on STEM education are Congress' priorities, then technological development, linking to the needs of business and industry, and education may become NASA's primary goals.

In: Space Policy and Its Ramifications
Editor: John P. Ramos

ISBN: 978-1-61761-555-9

Chapter 1

ADVANCE SPACE TECHNOLOGY IN THE LINEAR TRANSPORTATION SYSTEM FOR SAFETY AND SECURITY PURPOSES

Rustam B. Rustamov[1], **Saida E. Salahova**[2], **Sabina N. Hasanova**[3], **Maral H. Zeynalova**[4] **and Elman Aleskerov**[5]*

[1]Institute of Physics, Azerbaijan National Academy of Sciences, Baku, Azerbaijan
[2]Institute for Space Research of Natural Resources, Baku, Azerbaijan
[3]Architecture and Construction University/ENCOTEC LLC, Baku, Azerbaijan
[4]Institute of Botany, Azerbaijan National Academy of Sciences, Baku, Azerbaijan
[5]INTEGRIS, Baku, Azerbaijan

ABSTRACT

Safe pipeline laying-out and transportation of energy resources are a major concern for the public and the pipeline industry.

There are several aspects for support of critical infrastructure protection using the remote sensing for oil and gas pipeline safety purposes. It is an improvement of the transportation security, social and economical benefits, natural disaster, risk to property and human life etc.

Different countries use of different regulations and instructions for management of actions and the monitor of pipelines for transportation of natural oil and gas. Irrespective of details of individual requirements, it is an interest of any owner of the pipeline to support the pipeline in working condition and protect them effectively against the damage caused by the third parties. The general methods of monitoring most widely used for pipelines of transportation of natural oil and gas include patrolling among the route of the pipeline and remote observation from the appropriate height. For realization of monitoring of pipelines it is widely used small planes and helicopters. Though these methods are the high guarantor of safety of oil and gas pipelines but in this case the price

* Corresponding author: Email: r_rustamov@hotmail.com

is also enough high. Significant progress in the field of technology of high resolution remote sensing data used as plane and satellite onboard sensors and reduction of the cost due to increase of a memory capacity of computer technology and expansion of an opportunity of data processing technology was created with favorable conditions of application of this technology in safety and security issues of pipelines in transportations of oil and gas.

Today opportunity of high resolution of space imagery use creates a positive environment on application of space technology for monitoring of integrated systems in different areas of industry and commercial purposes. They are an airborne, unmanned aerial vehicles (UAV) and satellite Synthetic Aperture Radar and LIDAR multi-spectral and hyper-spectral sensors.

One of the advance methods of feasibility study of appropriate services for monitoring of integration systems is based on remote sensing data and Geographic Information System (GIS) developments. Objective of this approach is to improve safety, security aspects of integration systems, reduce survey costs and improve transportation and transmission efficiency through an increased monitoring frequency.

Earth observation for investigation requires very high resolution optical and in most cases radar sensors. A very high resolution is here defined as a resolution of 1 meter and better. This scale is required to allow the detection of targets. A very high resolution satellite observation, formerly limited to national and strictly classified reconnaissance tasks, has become a commercial business in the recent years. Affected by satellites losses and pioneered with Space Imaging's IKONOS satellite, better than 1 meter resolution optical data is now available for science and commercial use.

The use of advance technology makes available to enhance planning, design, management, operation and maintenance of the linear infrastructure. Aerial and satellite data based on the remote sensing methods with further integration to GIS technologies represent the area of rapid responds and developments which can be leveraged to assist the linear systems risk assessment to assure the safety of linear infrastructures. Industrial and scientific achievements in satellite remote sensing methods and appropriate data processing techniques are opening a new wide technological opportunities and availabilities to develop an enhanced capability to accomplish the linear systems mapping and safety needs of the oil and gas transportation. Combination of those technologies with GIS has a valuable and significant potential for further application to a huge of cross cutting system of security issues.

Keywords: Remote sensing, space image and data, resolution of space images sensor, image processing, security, safety, pipeline.

INTRODUCTION

Safe pipeline transportation of energy resources is a major concern for the public and the pipeline industry. From the first stage a pipeline has been built and buried, geologic and natural hazards, corrosions and third-party damages all pose cumulative internal and environmental risks to the pipeline's integrity. As today's pipeline engineering and operation become more reliant on geospatial data for safety in the pipeline's life cycle of design, construction, maintenance, and emergency response of pipeline facilities, the rapid and cost-affordable acquisition of terrain data along the pipeline corridor becomes increasingly critical.

Security and reliability oil and gas operational and transportation infrastructures are one of the significant technological elements of energy safety maintenance. Geographical scope of the transportation infrastructure is a huge - from Western Siberia and Central Asia up to the western territory of the European Union and from Northern Europe and adjoining Arctic regions up to Northern Africa and Middle East. The general aspects of the security and safety of oil and gas pipeline can be identified as the following:

- Geohazard studies;
- Environmental monitoring;
- Geographic Information Systems (GIS) applications;
- General risk assessment;
- Environmental Impact Assessment (EIA) studies;
- Environmental Management and Monitoring Plan (EMMP) and other related plans;
- Social Impact Assessment (SIA) studies.

The Baku-Tbilisi-Ceyhan (BTC) project is a $3 billion investment to unlock a vast store of energy from the Caspian Sea by providing a new crude oil pipeline from Azerbaijan, through Georgia, to Turkey for onward delivery to world markets. Traversing 1,768 km of often remote and challenging terrain, the BTC pipeline is able to transport up to one million barrels of crude oil per day from a cluster of discoveries in the Caspian Sea, known collectively as the Azeri, Chirag, deepwater Gunashli (ACG) field. By creating the first direct pipeline link between the landlocked Caspian Sea and the Mediterranean, the BTC project brings a positive economic advantage to the region and avoid increasing oil traffic through the vulnerable Turkish Straits. A programme of social and environmental investment is ensuring that the people of the three host nations also share in the benefits (Figure 1).

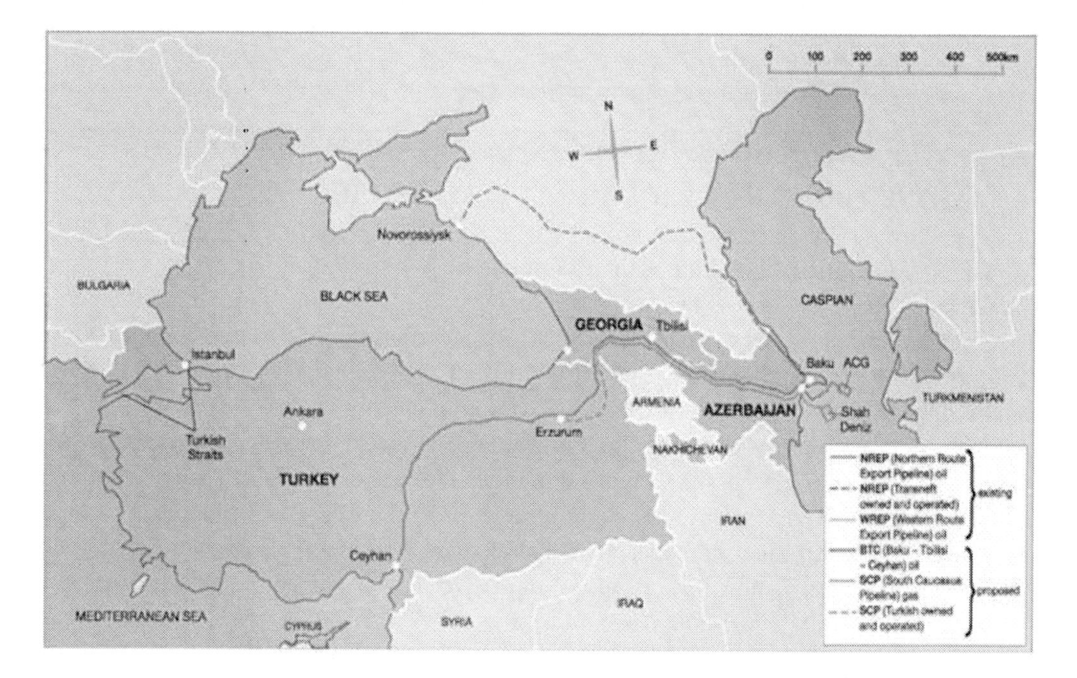

Figure 1. A map of the oil and gas pipeline network in the Caspian Sea region

The similarity of problem existing for oil and gas transportation can be identified for other regions around the World. It is important to undertake a necessity of application of available methods as as well as space technology for the issues of security and safety of the linear infrastructures.

Today the safe energy supply is one of the major factor in the stability of the economy and an important driver in international policy. Besides carrying for good relations with those regions of the world, where the prime energy is exploited, the transport of this energy to the recipients is attracting more and more attention. This issue starts from the reliable relations between country participants related to the oil and gas transportations as well as the safety supply of energy sources up to the end point.

In the meantime the cost of the transport is one the major factor and contributes – in the case of natural gas – significantly to the end user price. The security of pipelines is a matter of concern long before issues of "homeland security" appeared on the political agenda. In most countries involved for energy transportation, third party interference is mainly caused accidentally by construction activities. Nevertheless it is needed to improve and undertake a new concern in security against terrorist acts promoted the use of new technologies to reconsider the traditional observation and surveillance tasks.

In addition a new concerns and new military demands have enlarged the availability of spaceborne capabilities, which are suitable for a regular observation. New image analysis approaches would allow to automate this surveillance to a large degree.

However, the current ways of operating satellites and trying to make commercial business with selling the space data are hardly meeting the technical and financial demands of the pipeline operators in the regular survey and observations of a long pipeline tracks. It demands an approaches and ideas for successful implementation of the security and safety transportation of oil and gas.

1. Earth Observation as an Element of Pipeline Monitoring for Linear System Security and Safety

As per identified the European Commission White Paper on Space (European Commission 2003) describes the European strategy on developing independent access to space technology and operational capabilities. Besides defining the framework for Global Monitoring for Environment and Security (GMES) the Commission included a chapter entitled "Space as a contribution to the CFSP (Common Foreign and Security Policy), the ESDP (European Security and Defense Policy) and to the anticipation and monitoring of humanitarian crises". Therein the Commission urges the reinforcement of space technologies in support of security and defense policy requirements. While deliberately addressing military and dual use requirements, the paper and several other documents (European Commission, 2004; Gubert 2004; Dillon, 2005) also emphasize the need to protect critical linear infrastructure [1, 2].

In this approach the transportation lines for energy and other critical goods are especially mentioned.

For the success of the initial R&D projects, managed by the European Commission in the domain of Space and Security, a "Network of Excellence" on "Global Monitoring Of Security and Stability" (GMOSS) was launched in 2004, which aims to integrate Europe's civil security research so as to acquire and nourish the autonomous knowledge and expertise in Europe for research and applications for security based on satellite image information. GMOSS addresses generic methods and algorithms for automated image interpretation as well as application scenarios such as border control, infrastructure mapping and nuclear treaty monitoring. Managed by the German Space Agency (DLR), GMOSS integrates about 25 organizations and companies in Europe and run for four years (intelligence.jrc.cec.eu/gmoss). In GMOSS, specific work packages are concerned with the state-of-the-art in pipeline security as well as in technological approaches in automated feature detection available in Europe [3].

The success of the pipeline monitoring scenarios can be implemented if sufficient supply from spaceborne data will be available for reasonable economic circumstance and with certain technical performance. As it was identified a very high resolution is here defined as a resolution of 1 meter and better. Requirement for such resolution is required to allow the detection of pipeline interfering targets [4]. With insecure defense budgets in the future, commercial satellite data providers are also looking ahead to explore a new markets. Currently much smaller than the market for national defense customers, pipeline monitoring could well be such a market in the future. Behind the general need for image intelligence from space, three major drivers push the supply of very high resolution space image data.

The 25 member states of the European Union have declared space and security issues a major political and technological strategic task of the future. Under the driving program of the "Global Monitoring for Environment and Security" (GMES), especially the security issues are now being addressed by a growing fleet of European imaging satellites, which shall guarantee an independent access to critical and global geo-information. Besides the continuation of optical satellites in a higher resolution domain (e.g. the French Pleiades system), radar satellites with day and night and all weather imaging capabilities are on the European schedule (e.g. TerraSAR-X from Germany and Cosmo-Skymed from Italy). The European Space Agency (ESA) Sentinel fleet of satellites will complete the European capabilities, though their capabilities (i.e. especially the geometric resolution) are more aimed towards environmental and repetitive land and sea mapping. The high resolution national systems serve a dual-use market either by definition or by market demands. That means that all systems are also targeted to receive considerable revenue from non-military, commercial markets. Pipeline monitoring/planning is one such valuable emerging commercial market [5].

The use of a very high resolution space data will not only be satisfied by North American and European suppliers. Some emerging economies and even former developing countries have embarked on national Earth Observation capabilities, which are comparable and even better than the US and European systems. Except covering national security issues all systems are and will be available to apply within the framework of the international customers under commercial terms. As an example Indian high resolution satellites (e.g. the CartoSat series), Taiwanese (RocSat) and Thai (Theos) data can be incorporated in future operational infrastructure survey services. Interestingly, global data suppliers have positioned themselves to offer customers a range of different high resolution satellites: Space Imaging is the global distributor for the Indian Satellites, whereas SPOT Image – besides being appointed to distribute Pleiades data – has got the global distribution rights for the Taiwanese and Thai satellites.

1.1. Requirements for Pipeline Observation

Today, the share of natural gas in Europe's primary energy consumption is more than 20 %. This figure is expected to increase to about 30 % by 2030. The gas is transported via high pressure (up to 85 bar), large diameter (up to 1,2 m) steel pipelines from sources within and outside western Europe. The 450,000 km of high pressure pipelines form the backbones of the international gas transport, starting at the site of production and ending at certain national and regional distribution points, from where the gas is transferred to distributors serving the individual customer. Two thirds of European gas supplies are from West European production, whereas Russia is the main supplier in interregional gas trade. Natural gas and its production and transportation infrastructure form an economic and security aspect for the European society as well as for the involved gas providers. In comparison, the oil transport pipeline network is of rather reduced size in length, with the NATO fuel pipeline network of some 10,000 km forming the larger part (European Space Agency, 2003).

There are huge aspects which should be undertaken in safety and security issues of the pipeline infrastructure. Safeguarding the integrity of the pipelines was a point of concern for the pipeline operators long before the advent of the theme "homeland security" on the political agenda. However the primary concern in Europe was and still is not criminal acts to the pipeline infrastructure, but mostly accidental interference with the pipes. Nearly all high pressure gas pipelines in Europe are buried in the ground and may only be recognized by sign-poles indicating the underground presence of a gas pipeline about every 500 m. Though, underground pipelines have a lot of technical and environmental advantages. But they can also be a target of certain manmade and natural risks.

Construction activities such as laying of cables, building of houses and deep-plough farming could scratch the steel of the pipeline and therewith damage its anti-corrosion protection shell. In worst cases such construction activity could cause a significant gas escape. Such "third party interference" events contribute the biggest threat to the pipeline network integrity on the European scale. The current measure against these threats is to observe the pipeline either by walking (patrolling) along the lines (mainly in urban areas or easily possible patrolling) or by regularly flying the line with helicopters or Unmanned Vehicles or fixed-wing aircraft (for most of the pipeline network). Whilst the pilot is often guided by electronic maps based on developed GIS and satellite navigation system GPS to fly the route, the observation is left to the naked eye of the human observer. Only in a few cases digital pictures or`videos are taken. The frequency of such flight surveys varies from country-to country. It depends some of factors for instance complexity of the studied areas, relief and so on. But the survey can be assumed to be performed about every two months up to every two weeks. Giving the extend of the pipeline infrastructure and the potential economic losses and even threats to lives the regular observation and survey of the pipeline is a costly but necessary task in the point of view of the security and safety issues. The concern of the studies being discussed is to demonstrate the capabilities of a new high resolution earth observation sensors, linked with automated image analysis/feature recognition and Geographic Information Systems (GIS) for substituting the helicopter or Unmanned Vehicles and fixed-wing aircraft based survey with space borne assets.

In addition to manmade threats, also geologic risks could harm the integrity of the pipeline. Motion of the surface which happens regularly and therewith the motion of the buried pipeline – could eventually lead to a damage of a pipeline. These motions can be induced by underground mining activities, land slides and seismic events, just to name a few. In many cases, the exposition of land areas to surface motion is already known at the time of pipeline construction. Either avoiding the problematic areas or measuring the soil motion with ground based devices is an appropriate measure in these cases. However the observation of large areas is costly. Space based radar interferometry delivers a tool for a frequent and large area surface motion observation. Hence it become very useful is to study a radar interferometry as an important instrument for warning measure for protecting pipeline against geotectonic risks.

Finally, the study efforts also tried to detect eventual small leakages of pipelines, by measuring the methane either directly or by the environmental effect methane is imposing to the vegetation. Whilst the direct detection studies have been performed with Laser based measurements (LIDAR) the effect on the vegetation is primarily being measured with high resolution multispectral imagers.

1.2. Approaches to Pipeline Monitoring and the Resulting Benefits

Research activities and application needs and efforts are developing cost-effective ways to enhance pipeline integrity, inspection and monitoring, as well as new tools and techniques for managing the risk involved in pipeline operations. Some of these efforts have investigated the use of satellite-based technology for pipeline protection. It has been identified the potential for satellite imagery to detect significant slope motion and ground movements that could threaten nearby pipelines with a less expensive means.

Pipelines and their associated facilities pose potential environmental pollution risk that can also be monitored by a variety of remote sensing systems. Pipelines are required to meet all environmental pollution risk hazardous waste management standards. Hazardous wastes related to the transport process of oil and gas in pipelines is fully covered by environmental pollution risk regulations for the control of hazardous material.

New technologies based on application of advance space sciences for energy and transportation monitoring can provide following services:

- More Frequent Surveillance;
- Earlier Detection;
- Enhanced and automated incident detection/alerts;
- Video capture and archiving;
- Deterrent to threats (unpredictable, random schedule);
- Reduced industry's cost repair damage;
- Enhanced Right of Ways Security;
- Reduced Risks to Pilot/Public;
- Reduced 3rd Party Damage.

Foregoing indicated issues have been announced by the NASA's Earth Science Mission as " ...to conduct aeronautical and space activities so as to contribute materially to ...the expansion of human knowledge of the Earth and of phenomena in the atmosphere and space".

For this reason guides NASA's development and infusion of pioneering technologies and systems undertake:

- Observe, understand and model the Earth System;
- Detect how it is changing;
- Better predict change; and
- Understand the consequences of life on Earth.

Advances in information systems, satellites imaging systems and improved software technologies have led to opportunities for a new level of information products from remote sensed data. The integration of these new products into existing response systems can provide a wide range of analysis tools and information products that were not possible before. Using the higher resolution imagery and change detection analysis pipeline situational awareness and damage assessment can be conducted rapidly and accurately. Power utility infrastructure and system wide evaluations over a broad area could also be addressed using remote sensing data sources. All of those information products can be useful in the response, recovery and rehabilitation phases of infrastructure management preparedness.

1.3. The Main and Necessary Actions

The main and necessary actions which should be executed for safety purposes are a detailed monitoring and review of the pipeline operator requirements and the currently available remote sensing technologies that could meet these. From these actions a best practice approach will be evolved, identifying suitable approaches that could be considered now and those that will be available in the foreseeable future.

For investigation of these fantastic approaches, several of the partners will provide state-of-the-art monitoring equipment that will be flown over pipeline test sites to assess the performance of new techniques and the processing of multisensor data sets.

Outcomes of those investigations will be compared with imagery collected from current satellite instrumentation to enable the technology limits of the various approaches to be compared. The pipeline operator companies will provide access to their pipeline right-of-ways and will develop models for the risk and cost benefits of the various approaches considered.

It is necessary to indicate that a monitoring of pipelines and pipeline rights-of-way (RoW) is more important due to the following significant issues:

- Risk of terrorist attack on critical national infrastructure;
- A catastrophic event could jeopardize the public and the environment;
- Detection of ground disturbances, equipment/spills under the tree canopy, and crude oil leaks below snow cover;
- $$$ Millions spent to repair damage;
- Reputation of energy industry;
- License to Operate.

The requirements for the structure of a computer based management system that will collect the information, undertake the processing of the data and disseminate it to the end user pipeline operators will be established and a pilot demonstration system developed.

1.4. Expected Contributions and a New Developments and Achievements

The major advantage of used means are anticipated by replacing conventional monitoring techniques with remote surveillance sensing and employing advanced data processing of imagery to supply feature identification with improved response times compared with current practice.

It has to be taken into account that there are three key aspects which hold particular interest for pipeline operators: the priority is to assess how satellite monitoring from space could be used to identify and anticipate third party interference, such as tractors and large earth moving equipment, to the pipeline network.

The last third aspect is oil and gas leakage. Current pipeline monitoring is undertaken routinely using a number of techniques, the most common being as it called "flying-the-line" using helicopters or fixed wing aircraft with trained observers on board. These routine and regular flights perform a key role but they are enough expensive and bear the inherent safety concerns of all helicopter flights.

For successful implementation of pipeline safety and security aims it has to be involved the development of mathematical modeling software, image processing software and software to extract and select features from needed data collected from a wide range of detector instrumentation. A remote monitoring system would be transmitted images to Earth from satellites in space, following the route of a pipeline and identified key threats to pipeline safety and security. The final information will need to be presented to the end user in a simple and manageable format that will enable the recipient to respond to incidents in a measured way, without the probability of false alarms causing the quality, safety or integrity of the system being called into question.

Transportation of oil and gas demands permanent monitoring for maintenance of reliable safety issues. Up to date used the methods of monitoring for safety infrastructures are weekly monitoring on foot or horse, use of a various kind of vehicles, including helicopters with the purpose of exception of possible incident by the third party.

Recently thank to the successful development of the high resolution space technology this technology being applied as a more reliable safety system for oil and gas transportation. There are the successes in the field of various SAR systems, optical sensors, automatic methods of processing and management of pipelines systems.

The definition of the most effective method for providing suitable and successful transportation of oil and gas through the pipelines and solution of problems related to the ecology of environment is the main requirement aspect of oil and gas safety transportation. Annually collects the statistical data which presents information on incidents as a result of intervention of the third party, ground landslides or spillage of methane. There is the following approach for preliminary testing and demonstration of methane and oil spill detection:

- Direct Detection of CH4 gas above snow and ice;
- Direct detection of organic liquids on snow and ice;
- Direct detection of CH4 gas above water;
- Direct Detection of organic fluid spill on land using Hyperspectral RMOTC 2004 imagery;
- Direct Detection of CH4 gas using Hyperspectral RMOTC 2004 imagery;
- Detection of plant stress induced by underground CH4 leaks using Hyperspectral RMOTC 2004 imagery;
- Habitat mapping for baseline and change detection using Hyperspectral RMOTC 2004 imagery.

The purpose of investigations for finding out a new way of solution of problem is development of a new approach of management for the infrastructure with use of satellite monitoring system. It can be undertaken:

- Selecting, advancing, testing and demonstrating imaging and sensor technologies to enable automatic detection, and overcome adverse monitoring phenomenon such as fog, rain or darkness;
- Improving data management, integration and interpretation systems to enable remote capture and processing of information, threat detection, automated alerts, archiving, and human-systems integration.

Improving airborne detection capabilities (manned to unmanned airborne systems (UAS)) to enable safe, cost effective and frequent Right of Way (RoW) surveillance and monitoring. Additionally, a NASA-Pipeline Industry-DOT/PHMSA collaboration can build on NASA Ames' long proven and unique history with the:

- development of remote sensing and airborne systems in conjunction with the commercial sector;
- detailed development of requirements and well coordinated deployment of assets in conjunction with regulatory agencies (FAA, etc.); and
- experience and expertise in bridging the needs and capabilities of the private and public sectors, pushing further technological development while defining and addressing project specific issues in a manner consistent with regulatory requirements.

This approach allows operators of pipelines to carry out permanent monitoring indicated of pipeline status in any weather conditions and day.

The typical reasons of interaction of thirds of party with pipelines characterize works on digging, trickling or excavation around of pipeline and the use in this case of various mechanized machines. Those systems of risk can be determined by three stages:

- Change detection;
- Categorization of changes into priorities;
- Classification of categorized changes into hazards as digger, excavation, tree, etc.

The last item of this approach can provide the security and safety issues of pipelines related to acts of terrorism.

Certainly general issue of oil and gas pipeline safety includes aspects of the natural disaster and problems related to the environment.

2. CHANGE DETECTION, THEIR CLASSIFICATION AND MONITORING OF LINEAR SYSTEMS

Detection of change characterized the following:

- The first approach is an algorithm of detection of the post classification change where the changes found out on the based of classification results of the ground surface. Changes occur, if a surface of the ground between two images differs from each other. An important feature in classification of the ground surface is the normalized difference vegetation index (NDVI) which is indicating relative quantitative of vegetation;
- The second approach on the basis of pixel includes differences between algorithms of detection of post reclassification changes. The first algorithm calculates temporal scalar product (TSP). The second algorithm determines the changes in vegetation calculating NDVI. TSP and NDVI merge as one layer;
- The third algorithm is a statistical and result of the false information. Both approaches are used in optical detection change.

By priorities of changes are classified of possible damages. Because of the present possible resolution of space images within the limits of 0.5 - 0.8 meters the decision of the problem is not so simple. Complexity of the specification of classes demands their restriction in the following definitions:

- mechanisms of the small dimensions, small excavation;
- mechanisms of the small dimensions, small road works;
- mechanisms of the big dimensions, the big excavation, planting of trees;
- mechanisms of the big dimensions, the big road works;
- work on a surface of the ground, planting of woods;
- works on preparations and transportations of wood;
- destruction of constructions and buildings;
- disappearance of water and water reservoir;
- new formations of water;
- other damages.

The technology of processing of the element of pixel of the image frequently does not allow realizing to recognition of object even when it is visible by human visual survey. It is caused that classification of the image on the basis of pixel uses the spectral information presented in the digital form within one or more spectral channels. In this case try's to classify each pixel, mainly on the basis of this spectral information. During the data processing the following concept is taking account: in this case this provides not only pixels of the image as well as an importance of objects of the image and mutual relation between them. The hierarchical network of object of the image is used with the purpose of selection and combination appropriate scales for classification of the image.

2.1. Monitoring of the Ground Surface Movement over the Pipeline Corridor

Today both optical and radar systems are operated on airborne as well as space-born platforms [6]. Spaceborn systems presently provide a geometric resolution on the ground of up to 0.6m x 0.6m for optical and 10m x 10m for radar systems allowing imaging of strips with a width of 11km and 50km, respectively. Optical sensors are hindered by cloud cover which from an operational point of view constitutes a substantial limitation of their use in pipeline monitoring. Conversely radar sensors can be operated regardless of weather and light conditions and therefore have a much higher rate of availability.

Conceptually monitoring the integration system is structured into four main system components:

- the Pipeline Operator System (POS), which is the part of the monitoring system which is used by the pipeline operator for delivery and handling of alarms and for specifying the monitoring characteristics for different parts of the pipeline network;
- the Pipeline Information Management System (PIMS), which stores all relevant information on the pipeline network, the environment around it, and the integrity monitoring and which provides analyses and scheduling functionalities for the pipeline operator. The PIMS also includes an alarm production system, which decides what hazards should be considered as alarms;
- the Hazard Extraction System (HES), which extracts the hazard report information out of the basic remote sensing imagery layers, using advanced image interpretation techniques;
- the imagery Collection System (ICS), which collects the required remote sensing imagery with a suit of both spaceborne and airborne platforms and different types of sensors, conform the monitoring priorities. In the ICS all these means are scheduled optimally conform the specified priorities of the pipeline operators and the weather and season conditions. Here also the data are pre-processed to remote sensing basic imagery layers.

The four components in principle can be independent of each other so that maximal flexibility exists. Also each system component in itself is set up as much as possible in a modular and flexible way. By implementation of this above mentioned issues, new technologies on sensors, platforms, data processing, data storage and transfer can be integrated and the system easily can be extended to other operators or areas.

The future integration systems monitoring scenarios can only be turned into reality if sufficient supply from spaceborne data will be available for reasonable economic conditions and with certain technical performance. Earth observation for appropriate investigation requires very high resolution optical and in most cases radar sensors. A very high resolution is here defined as a resolution of 1 meter and better. This scale is required to allow the detection of targets. A very high resolution satellite observation, formerly limited to national and strictly classified reconnaissance tasks, has become a commercial business in the recent years. However, national security users are still the basis for the commercial viability of that business. Affected by satellites losses and pioneered with Space Imaging's IKONOS satellite, better than 1 meter resolution optical data is now available for science and commercial use.

Development in high resolution optical systems have gone relatively fast during the last years. For spaceborne systems the last years several commercial satellites have come available with panchromatic resolutions of better than 1m and multispectral resolutions of better than 3m. Currently a subdivision can be made between four types of systems:

- High quality photogrammetric systems. Currently the first operational photogrammetric systems are coming to market. The resolution and geometric requirements of these systems are very high. Main applications are large scale mapping and DEM generation activities. The processing software for these systems is very sophisticated and almost fully automated;
- Multispectral scanning systems. Multispectral airborne scanners exist for about 20 years already and more and more operational systems come available. Different examples exist of multispectral sensors operating in the VIS/NIR region, both pushbroom scanners and matrix cameras. Also examples exist of operational sensors operating in the VIS, NIR and MIR region, e.g. the Daedalus (whiskbroom) scanners which operate with 10 to 20 channels. Important aspect for these systems is that the data processing software is strongly related to the system (sensor, radiometry, geometry) and therefore needed to be operational;
- Digital camera based systems. The quality of commercial digital camera's is increasing rapidly which means that these camera's are more and more suited for operational use. Especially the high end camera's provide good detail and radiometric sensitivity and are provide possibilities to influence the spectral band definition, quantization and automatic read out of the imagery. Examples exist of operational systems of one or a cluster of digital camera's including a highly automated and operational processing chain;
- Digital video based systems. Also the quality of digital video camera's has increased during the last years. Still the number of pixels and the sensitivity stays a limitation. It is expected that within a number of year very high resolution video camera's will enter the market with non interlaced images of 1000*1500. These might be suited very well for airborne monitoring systems. Also currently examples exist however of operational video based airborne monitoring systems. Even from false color video systems.

Interferential processing of satellite data Synthetic Aperture Radar (InSAR) allows to find out and analyze differences between two or a big images, registered on various orbital positions. As a result of such approach of solving of the problem enables to register changes of position of elements of the ground within the resolution of or millimeter dependence of application of used equipment and technology.

Presently application of InSAR allows analyzing the following basic forms:

- Standard InSAR: processing of differences by phase between individual pairs of the whole image reflected from the surface;
- Corner Reflector InSAR (CRInSAR): processing displacement between two or more images which are applied only for specially constructed reflectors;
- Persistent Scatter Interferometer (PSI): stacking of repeated images, that is 15 or more to select and use only returned back from natural constants, strong reflector signals.

(i) Standard InSAR. These radar systems are applied in cases where are lacked a high density of strong reflective signals. Similar systems are used in the area where displacement makes within the limits of centimeter and more. The time interval for registration of the image depends of the behavior of the surface investigated object. But the typical time interval makes from several weeks about one month. This technology allows realization of monitoring of objects such as growth of vegetation, processing of the agricultural ground, etc.
InSAR earthquake, etc. on the basis concerning cheap space images can be used as a good instrument covering a large area of image, for example slides of land within the big scale;

(ii) Corner Reflector InSAR (CRInSAR). Radar reflectors (reflectors of corner, or active receiver/response) can be used where the strong reflection of the signal of the satellite from separately taken point is provided. Use of similar systems during construction of pipelines allows of fixing slides of the land surface with a high accuracy. Application of such radars is very effective and suitable where probability of risk and danger are a high;
However CRInSAR has disadvantages for the point of view of the technical application. For operation of these radars the reflectors should be located in a distances from each other not less than hundreds meters that would be possible of differentiation of influence of atmospheric influence and at least one of reflectors should be in precisely fixed position playing a role of the reference reflector;

(iii) Persistent Scatter Interferometer. In the city surrounded with rocks where forms the strong reflections from objects realization of qualitative collection of space data has a significant importance. Gathering images of SAR, for example 15 or more it is possible to identify and also delete or select from process of the further processing all features being time (for example, vehicles, etc.) and to keep only necessary, requiring elements for the further analysis. Quantity of reflectors can be in the interval between 50 and 400 points for each kilometer square.

Thus, in SAR the problems caused by atmospheric influences, landslides and errors because of orbits of satellites are successfully enough solved. Such opportunity allows obtaining the high quality satellite data for the millimeter resolution. Opportunities of cover of a large area of the surface of the land make advantage of those systems than GPS in case of definition of the level of the place and positional problems.

3. Application of Remote Sensing and GIS Technologies for Pipeline Security Assessment

Advances in geospatial sensors, data analysis methods and communication technology present anew opportunities for users to increase productivity, reduce costs, facilitate innovation and create virtual collaborative environments for addressing the challenges of security improvement and risk reduction. Sensor developments include a new generation of high-resolution commercial satellites that will provide unique levels of accuracy in spatial, spectral and temporal attributes. An example is Digital Globe's commercial Quick Bird satellite with a spatial resolution of sixty one centimeters. An example of an image from this system is the open pit mining and pipeline facility in Thailand (Digital Globe 2003) shown is Figure 2. This image was taken by Quick Bird from 450 miles in space and provides a degree of detail not available previously to the civil community from space based monitoring of infrastructure on the ground.

In addition to the high resolution panchromatic imagery illustrated in Figure 2, there are a number of other commercial imagery products that are potentially applicable to pipeline transportation and power industry infrastructure. They include air borne and satellite radar, LIDAR, multi-spectral, and hyper-spectral sensors. Part of the challenge is matching the best sensor to the specific transportation related application. Visualization and advanced data analysis methods are also important capabilities. Automated change detection within a defined sector is one example of analysis capability that will assist in detection of unauthorized intrusion events. A specific application of these techniques to power distribution security is the detection of unauthorized intrusion onto pipeline right of ways. Pipelines often cover thousands of miles and are located in remote areas that are difficult and expensive to monitor. In one case study satellite imagery and target identification analysis is used to detect unauthorized intrusion onto a pipeline right-of-way in a remote area of Canada [7].

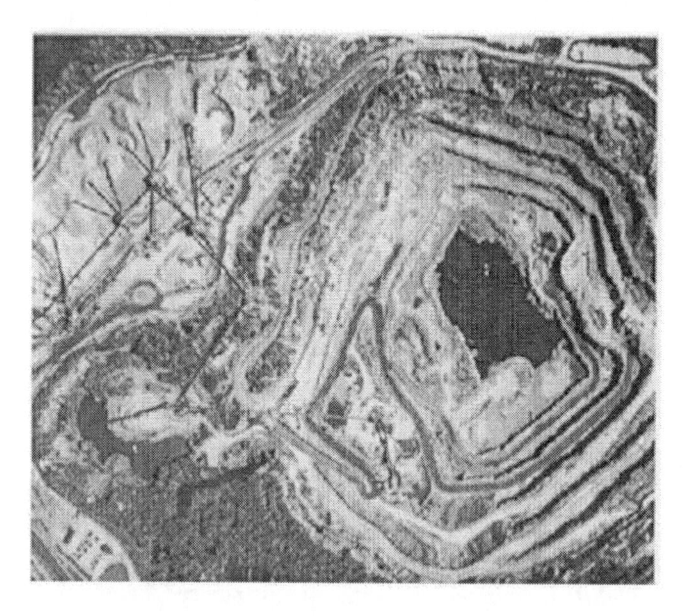

Figure 2. Digital Globe satellite image of Mining and pipeline system operation

The principle of optical sensors is that via an optical lens system the amount of sunlight as reflected by the earth is sensed. Normally only light in a specific spectral window is gathered, e.g. red, green or blue light. High resolution optical systems sense the light intensity for very small areas on the earth surface, in the order of 0.5 to several meters. Given this principle, optical sensors can be characterized by a number of important features:

1. Optical sensors correspond to the human eye in many aspects which makes interpretation easy;
2. Optical sensors can reach relatively high spatial resolutions. Resolutions of less than 0.5 meter from space and some centimeters from the air are possible. Limiting factors for the spatial resolution are the sensitivity of the detectors, the size of the lens or mirror and the reflection;
3. Optical sensors can only used in daytime. Since optical sensors make use of a passive light-source, the Sun;
4. Optical sensors are hindered by cloud cover which represents from an operational point of view an important limitation.

Developments in high resolution optical systems have moved relatively fast during recent years. For space-borne systems several commercial satellites have become available with panchromatic resolutions of 0.6m and 1.0 m and multispectral resolutions better than 3m. For airborne platforms a wide variety of optical digital scanners and cameras is available commercially. The principle of optical sensors is that via an optical lens system and a detector the amount of radiation incident from a target is sensed. A sensor gathers light in one or more specific spectral windows, e.g. red, green or blue light.

Panchromatic sensors. They have only one, relatively broad, spectral band, generally covering the region of 500nm to 700nm, resulting in a "black and white" image. Therefore the image contains little spectral (color) information. Differences between objects with different colors but the same intensities cannot be observed. On the other hand very high spatial resolution can be reached because of the wide spectral band, collecting relatively high energy levels. Sometimes the spectral band is extended up to 800nm. This gives the advantage that a significant amount of energy reflected by vegetation is included and also that the vegetation can clearly be recognized in the images [8].

Multispectral Sensors. These sensors contain a limited number of spectral bands (3 to 10). The spectral bands are defined by spectral filters which only pass light within a certain spectral region, generally with a bandwidth of 10nm to 100nm. Because of the narrow spectral bands, energy is limited and thus the spatial resolution of multispectral images is usually less that for panchromatic images. The most wide spread kind of multispectral sensors show three bands: green (550nm), red (650nm) and near infrared (850nm). These three bands are represented as false color image, in blue, green and red respectively. Some multispectral sensors do have more spectral bands.

Blue band (400nm). Combined with the green and red band this gives natural color images. Also this band contains information on the atmosphere so that it can be used for the atmospheric correction of the other spectral bands.

Bands in infrared (1.000nm – 15.000nm). While spectral bands in the near and middle infrared (900nm – 2.5nm) mainly contain information on vegetation, minerals and soil moisture, the spectral bands around 10.000nm are governed by the emitted thermal radiation of the surface and contain information about temperature and emissivity.

3.1. LIDAR Systems

Airborne LIDAR is an aircraft-mounted laser system designed to measure the 3-D coordinates of Earth's surface. It has been proven to be an effective technology for acquiring terrain surface data with high accuracy. The LIDAR may provide a supplemental technology to pipeline risk management to assure safety in design, construction, testing, operation, maintenance, and emergency response of pipeline facilities. It provides rapid 3-D data collection of long, linear objects such as pipeline corridors, roads, railway tracks, waterways, coastal zone or power lines (Fowler, 2000). It is easier to obtain many terrain parameters (e.g., slope) and to generate a 3D fly-through using these data. Since IDAR systems have a narrower swath width in comparison to many optical sensors, they are well suited for capturing high resolution long linear ground features in 3D, such as pipelines.

3.2. Terrain Analysis

When the LIDAR is combined with a digital photograph, the client has the added value of an image geo-referenced to the laser data set. By combining traditional photogrammetric mapping services with advanced data collection and processing techniques, the new technology provides pipeline monitors with better information for solving problems and making decisions. LIDAR data also could facilitate the planning of new lines and deciding safe routes for placement of a pipeline by considering the terrain parameters such as slope.

LIDAR-topographic-mapping systems have considerable promise for producing high-resolution digital elevation models (DEMs). Satellite communications and GPS navigation systems are critical parts of LIDAR mapping systems. In one study the Topographic Engineering Center (TEC) acquired multiple LIDAR-topographic-mapping data for the purpose of producing high-resolution DEMs. These were evaluated and the results used to help develop criteria for future applications for floodplain mapping (Roper, June 1999).

However, processing of raw LIDAR data into useful and reliable DEMs is not yet mature. Unless exceptional effort is made to produce accurate LIDAR calibration, significant merging artifacts, and their associated errors, can occur. Another example is with merging artifacts that can presently be found in Houston Advanced Research Center (HARC) data.

Merging artifacts in LIDAR data are generally due to miss-calibration of the LIDAR sensor or residual GPS errors. Both kinds of errors can be described by mathematical models and corrected. Because data within overlap regions should match, the parameters of the mathematical models can be calculated. Subsequently, the data can be adjusted so that the errors are minimized. Least-squares estimation techniques have proven to be efficient, accurate, and reliable for this purpose. The overall process is a least squares model-based merging.

Airborne LIDAR systems are capable of precise platform-to-ground ranging that produce decimeter-level height accuracy and meter-level post spacing topographic information of the surface. Some systems are capable of return pulse waveform digitization, which provide precise ground surface and vegetation height and volume measurements. Laser altimeter measurements are acquired along profiles or swaths up to a few hundred meters wide using small aircraft (Schnick, 2001). These swaths could replace channel and pipeline right of way cross-sections (currently obtained via costly ground-based surveys) required by flood backwater models, which are used to predict floodplain extent.

A portion of the Jet Propulsion Laboratory (JPL) LIDAR data collection effort was over the levee and pipeline systems of the Sacramento River valley near Sacramento, California. A segment of this DEM data set in the Meling Orange Grove area of California (Jorgensen, 1998) is shown in Figure 3. The three dimensional data bases that can be created from LIDAR sensor data can be viewed from different orientations and in a virtual fly-thru mode. The fly-thru capability allows the viewer to experience the terrain in a way that provides unique information about the terrain and a better understanding of the conditions on the ground.

3.3. Thermal Infrared Remote Sensing

Temperature differences between the pipeline and the environment, whether the lines are buried or above ground is another detection opportunity. The principle of thermal infrared imagery is that the detected objects produce maximum temperature difference against the background. In addition the weather condition at that time and the possible interference factors on the ground must be taken into consideration.

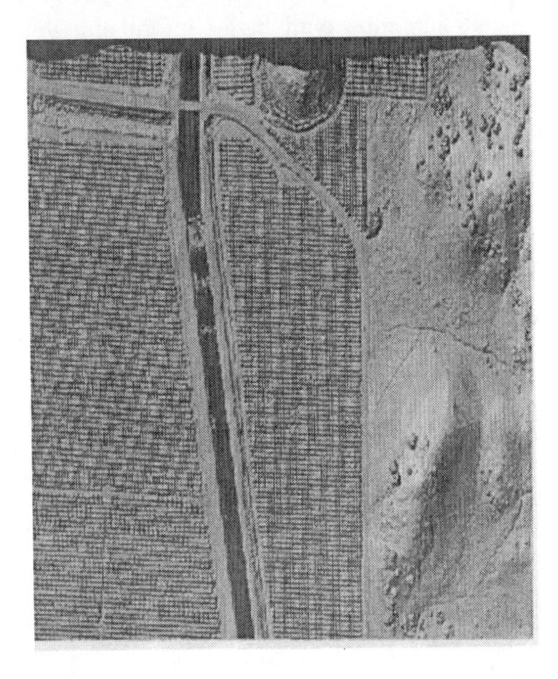

Figure 3. DEM generated from LIDAR data of orange grove, integration channel and pipeline

The thermal conduction caused by underground pipes usually forms temperature anomalies on the ground that often appeared as a difference in temperature. Imaging time: Imaging at night is the presupposition for the thermal target detection with the surface as background, because the ground temperature-increasing phenomenon caused by sun radiation is removed. Test data have shown that for high latitude area, the ground surface temperature tends to be stable at 2000 hours at night to 0600 hours the next morning at this season. The maximum difference in temperature usually appears just before dawn (Jensen, 2000).

The mathematical expression for object resolution is:

$$d = \beta H \quad (1)$$

Where:
d = Object resolution (m)
β = the transient view angle for scanning (2.5m rad.)
H = imaging flight relative altitude (m)

It is shown in the expression (1) that the object resolution is inversely proportional to the flight altitude, i.e. the lower flight altitude, the higher. Object resolution. Field test data have shown that the temperature anomaly width appeared on the ground for the petroleum pipe was usually 1-1.5m at night. Considering the safety of aircraft flying at night, pilots will often use 600m as their imaging flight altitude. The ground resolution (d) is in the range of 1.5m. So, the pipe anomalies easily detected [9].

Typical instrumentation is a DS-1230 quantitative double-channel infrared scanner with two blackbodies as reference source to control the upper and lower limitation of the received temperature signal. According to the ground test data and the signal in the flight, usual set limits would be BB2=-2°C (the upper limit temperature), BB1=-1.5°C (the lower limit temperature) for imaging. In this way, the radiant temperature from different objects on the ground was in the range. The upper limit of the normal temperature range is a little higher, in order to identify the pipe anomalies against the relatively dark background of the images.

Recent tests have successfully proven that the airborne thermal infrared remote is a new and efficient technology for petroleum pipe detection, and can be used in production. The selection of imaging season, imaging time, flight altitude and the normal temperature range for the thermal infrared remote sensing investigation are reasonable and the effectiveness of test were good. The advantage of airborne thermal infrared remote sensing for petroleum pipe detection is: good visualization, easily positioned and convenience for comparison and because it is not limited by access its cost is lower than the other ground geophysical methods.

3.4. Security and Safety Issues of Designed Engineering Infrastructure by Means of Space Technology

Today in many spheres of our life is successfully use of advances of space technology. One of the main aspects of this application is preliminary monitoring of the area with application of space technology undertaken for a wide areas of engineering developments.

Taking into account of presented approach it can be possible to create a positive environment for making highly accuracy decisions based on information received on processed satellite data. It provides spatial solutions to many fields of engineering such as transportation, water resources, facilities management, urban planning, construction and E-business [10]. At the same time it opens a possibility to consider needed natural factors like landslides, flooding, seismic issues etc. It is required to inquire of density of other engineering infrastructures and facilities in the designed area where space technology also can be play a significant place in problem solving.

The use of Remote Sensing (RS) methods and integration of received data into the Geographic Information System (GIS) is a suitable instrument in the engineering activities. GIS technologies have the potential to solve space related problems of construction industry involving complex visualization, integration of information, route planning, E-commerce, cost estimation, etc. GIS activities may be grouped into spatial and attribute data management, data display, data exploration, data analysis and modeling. The spatial and non-spatial data in GIS are synchronized so that both can be quarried, analyzed and displayed. Spatial data is related to the geometry of features, while attribute data stored in the tabular form describe the characteristics of different features of a layer in GIS. Each row of table represents a feature, while column represents characteristic of features. The intersection of a column and row shows a value of particular characteristic of that feature (Figure 4). GIS uses vector and raster data models to represent the spatial features.

The vector data model uses points and their x, y coordinates to construct spatial features (points, lines and areas). The features are treated as discrete object in the space. The raster data model uses a grid to represent spatial variations of features. Each cell of grid has a value that corresponds to the characteristic of a spatial feature at that location. Raster data is well suited to represent continuous spatial features [11].

Figure 5 illustrates all steps of processes of security and safety aspects of engineering facility starting from engineering design up to development of GIS database for appropriate state authorities to be needed for decision making.

The planning map of selected area which was a part of layer of the GIS has been developed for the aim of suitable location of engineering facility in the area and most convenient integration of the existed communication systems into the designed engineering facility (Figure 6).

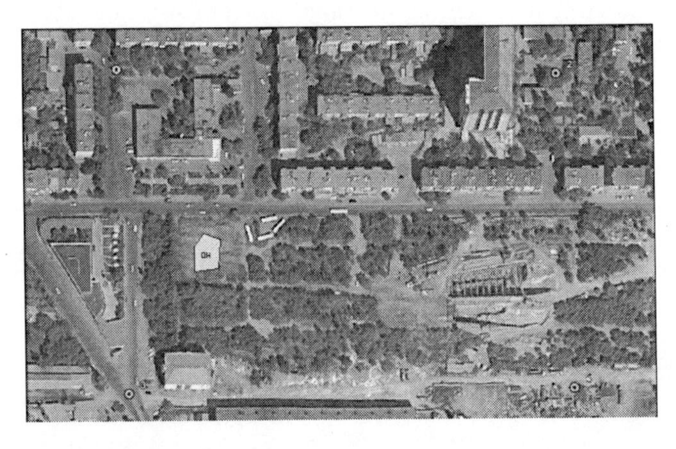

Figure 4. Space image of IKONOS with spatial resolution 1m for 2009

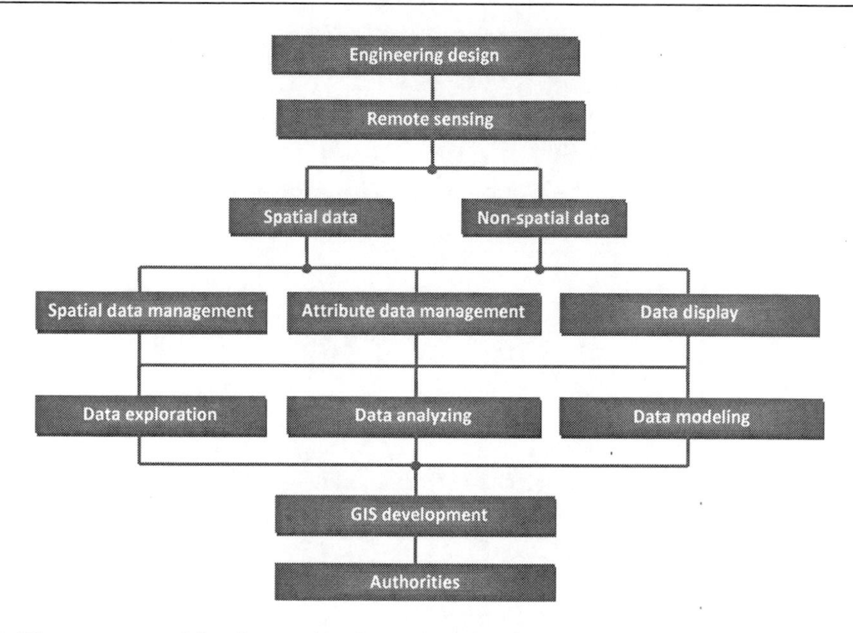

Figure 5. The processes of data integration for engineering facility security and safety issues

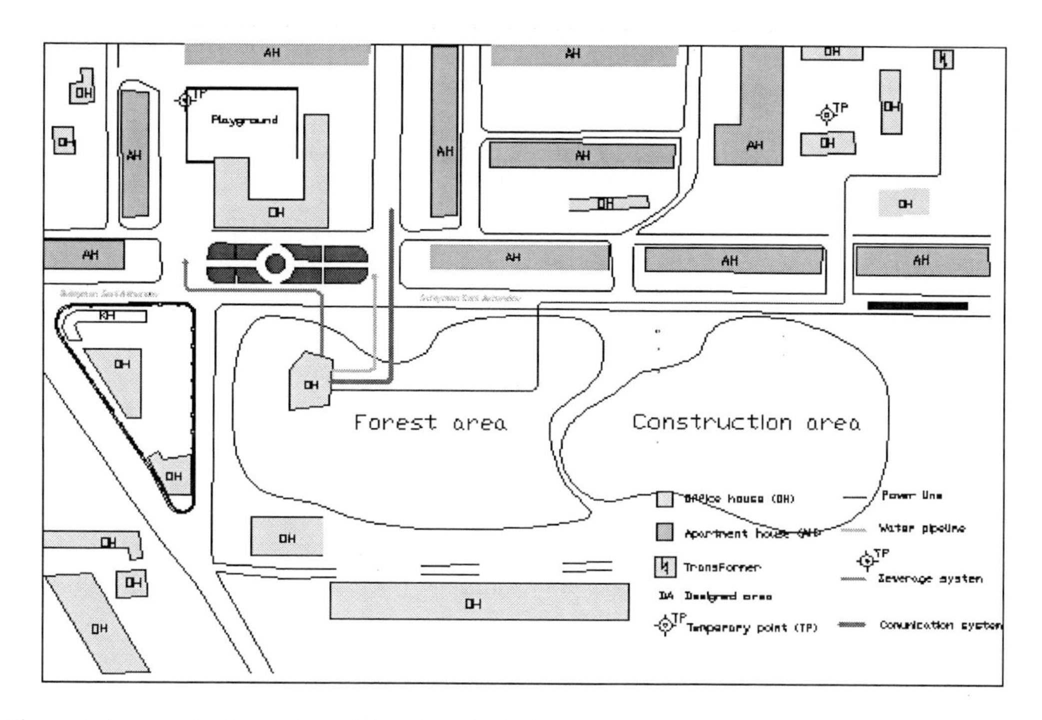

Figure 6. Communication systems integration design layout

Except advance use of space technology for selection of most suitable adaptation of communication systems there is an advantage of application of space technology for preparedness and forecasting all possible damages to the designed engineering facility. It was undertaken a design of engineering facility within the precincts of city as an example of maintaining of security and safety systems based on space technology applications (Figure 7).

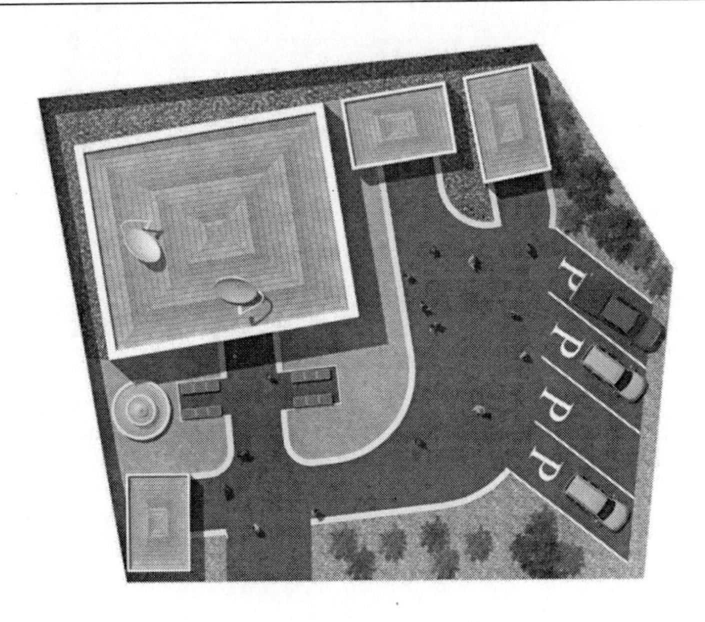

Figure 7. Architectural design of the engineering facility

As per identified GIS is a computer tool for capturing, storing, quarrying, analyzing and displaying the geographic information can be successfully used for security and safety targets of linear infrastructures in the complicated areas in point of accessibility, natural disaster and other possible issues. As GIS is one of the fast emerging fields being utilized in various engineering projects, its complete potential to the construction industry has not been realized yet. GIS improve the construction planning and design efficiency by integration of spatial and attribute information in a single environment.

4. Alternative Technologies to Spaceborne Missions

As per obviously accepted the observation of elongated corridors, may it be land borders, highways or pipeline tracks, are ideally suited for carriers, which are dedicated to fly along just this corridor. Satellites cross these corridors (which are mostly in east-west direction if European major pipelines are concerned) regularly in north – south direction, which means that only a fraction of their operational time could be used to monitor linear infrastructures constructed for oil and gas transportation aims. At the same time the operational costs of dedicated and manned aerial vehicles (helicopters and small planes) have forced the pipeline operators to consider possible alternatives for the safety and security purposes.

This status of use of civil unmanned aerial vehicles (UAVs) opens a huge and wide of opportunities and sa great potential to contribute to many surveillance applications where pipeline monitoring is also one of the success applications for safety and security issues. [12, 13]. There are some of reasons which make of use of UAV very attractive. One of them the civil UAVs can perform where manned flight is too dangerous, dirty and dull (so called D3 tasks). The other advantage presently smaller UAVs are affordable and can carry optimized and miniaturized sensor payloads.

However, the use of the Civil UAV is presently highly limited by the lack of regulations, standards and procedures necessary to operate the Civil UAV in a civil Air Traffic Control/Management (ATC/ATM) environment. The establishment of airworthiness and operational certification standards is necessary to open the airspace for Civil UAVs.

These airworthiness standards for UAV must ensure that the appropriate safety level (with respect to potential risks on the persons and the property on the ground) is met and that UAV gain public trust as well as social and political acceptance. For the case of monitoring and surveillance applications, a real cost saving factor would only be gained if the UAV is really operating fully automatically. Presently navigation technologies would indeed allow such operations, but at the cost of sophisticated automation and special concern in automatic Traffic Alert and Collision Avoidance System (TCAS) into UAVs.

Once these safety and guidance issues have been solved and operational certification procedures exist, Civil Unmanned Aerial Vehicles will be ready to operate in civil airspace and serve numerous civil applications including the potential of regular inspections of gas pipelines. The requirements with respect to sensor type and size for operational monitoring of gas pipelines, however, depend on the type of UAV. These can range from small and lightweight, low altitude UAVs with limited sensor weight and capacity, up to medium-size, mid-altitude systems with enough capacity to carry multisensor applications.

Available today observation platforms, such as UAVs will be an interesting and useful alternative, though today air traffic control regulations prevent UAVs to fly in civil airspace. One way to these problems is to have platforms which fly above this airspace.

4.1. Unmanned Airborne Vehicles

Unmanned airborne vehicles (UAVs) are an immerging category of airborne sensors that provide good resolution imagery at a lower cost than large scale manned airborne systems. Some of the UAVs, particularly military systems, have long range in-flight capabilities. Most of the small UAV systems, however, have limited range and flight time capability. For applications to pipeline monitoring larger UAV systems would probably best meet system wide coverage requirements. For critical segments of a pipeline system that is of more limited geographic area the small UAVs could be very effective. A number of Unmanned Aerial Vehicles (UAVs) presently exist, both domestically and internationally. Their payload weight carrying capability, their accommodations (volume, environment), their mission profile (altitude, range, duration) and their command, control and data acquisition capabilities vary significantly. Routine civil access to these various UAV assets is in an embryonic state and is only just now emerging.

A buildup of domestic UAV configurations, promoted by the Department of Defense (DOD) occurred in the late 1980s and well into the 90s. This occurred as the DOD sought UAVs to satisfy their mission unique surveillance requirements in either a Close Range, Short Range or Endurance category of vehicle. Close Range was defined to be within 50 kilometers, Short Range was defined as within 200 kilometers and Endurance as anything beyond. With the advent of newer technology and with the demonstrated performance of the UAVs provided to the DOD by industry, the Close and Short Range categories have since been combined, and a later separate Shipboard category has also been incorporated with them. The current classes or combination of these type vehicles are called the Tactical UAV, followed by the Endurance category.

For the potential civilian user of Unmanned Aerial Vehicles, the following categories are often used: LOCAL, REGIONAL, and ENDURANCE. Within these three categories of vehicles (LOCAL, REGIONAL, and ENDURANCE), approximately twenty-two companies domestically within the U.S. are or have been involved, and represent approximately forty-five different UAV configurations. The appearance of these UAVs and their known performance capabilities and payload accommodations vary considerably. They range in size from hand-held to runway-operated behemoths, whose payload weight capabilities range from a few pounds to 2000 pounds.

New capabilities projected for UAVs over the next 25 years include:

- Silent flight using fuel cells to supplant internal combustion engines in some systems.
- 60 percent gains in endurance due to increasingly efficient turbine engines.
- Rotorcraft capable of high speeds (400+ kts) or long endurance (24+ hrs) while retaining the ability to hover.
- Endurance UAVs serving as GPS pseudo-satellites and airborne communications nodes to provide theater and tactical users with better connectivity, clearer reception, and reduced vulnerability to jamming.
- Faster cruise missile targeting due to more precise terrain mapping by high altitude UAVs.
- Self-repairing, damage compensating, more survivable UAVs.
- Significantly speedier information availability through onboard real-time processing, higher data rates, and covert transmission.

UAVs will likely play a major role in the increasingly complex world of security and homeland defense that will evolve in the 21st century. There will be micro air vehicles as well as behemoths. UAVs will stay airborne for weeks or months and longer, fly at hypersonic speeds, sense data in revolutionary ways, and communicate their data at unprecedented rates. Challenges, such as providing an adequate command, control and communication infrastructure to capitalize on unmanned as well as manned operations, remain to be overcome. However, the decisions made now will lay the foundation for how far and how fast these advances are implemented. Only our imagination will limit the potential of UAVs in the 21st century.

Two possible UAV scenarios for a future pipeline inspection system

Civil unmanned aerial vehicles (UAVs) have a significant potential to contribute to the improvement wide areas of the life quality. The civil UAVs can perform where manned flight is too dangerous and impossible to use as well as expensive or monotonous. There are numerous applications where the use of Civil UAVs are highly preferable over manned flight, such as detecting storms and forest fires, or inspecting large infrastructures such as pipelines or electric power lines. However the use of the Civil UAV is presently highly limited by the lack of regulations, standards and procedures necessary to operate the Civil UAV in a civil

Air Traffic Control/Air Traffic Management (ATC/ATM) environment. The establishment of airworthiness and operational certification standards is necessary to open the airspace for Civil UAVs. [13, 14, 15].

These identified airworthiness standards for UAV which have to be undertaken for successful operation must ensure that the appropriate safety level (with respect to potential risks on the persons and the property on the ground) is met and that UAV gain public trust as well as social and political acceptance. The UAV airspace integration will bring a new operational aspects to the ATC/ATM community because of the differences in control, performance and flight objectives between Civil UAVs and general aviation.

Once the safety and guidance issues have been solved and operational certification procedures exist Civil Unmanned Aerial Vehicles will be ready to operate in civil airspace and serve numerous civil applications: UAVs seem to be suitable airborne platforms for the task of regular inspections of gas pipelines [16]. The requirements with respect to sensor type and size for operational monitoring of gas pipelines, however, depend on the type of UAV. For this purpose, two different scenarios for UAV based pipeline monitoring systems will be considered:

1. Small and lightweight, low-altitude UAVs with a limited sensor and weight capacity, e.g. the LUNA X-2000 system (Figure 11);
2. Medium-size, mid-altitude UAVs with a weight capacity sufficient for multi-sensor applications, e.g. the system EAGLE (Figure 12).

UAV scenario 1 – small and lightweight low-altitude UAV
A small lightweight low altitude UAV can be characterized by the following parameters:

Altitude:	low
Payload typical:	<25 kg
Endurance:	5-6 hrs

Such a UAV is operated in the uncontrolled (lower) airspace and therefore, requires appropriate sense-and-avoid technology to avoid collisions with obstacles such as buildings, electrical power lines or low flying objects (balloons). For the purpose of pipeline inspection, an altitude of ca. 100 m might be appropriate, which is generally below clouds. In this case, an optical/IR sensor system will be sufficient. Furthermore, high resolution can be achieved with even small optics. From the technical system and platform point of view, all components for UAV based pipeline inspection are available:

- The LUNA X-2000 platform has been operated successfully in Kosovo and Iraq;
- Lightweight optical/IR sensors are commercially available for airborne applications;
- On-board image pre-processing tools (hardware and software) exist;
- Data transmission to ground stations is no major technical problem;
- The image processing and feature extraction efforts are relatively simple, since the processing is restricted to one sensor.

However, safe automatic operation in uncontrolled low airspace has to be developed, which is a necessary condition to provide a feasible technical solution.

UAV scenario 2 – medium-size and medium-weight mid-altitude UAV
A typical MALE (Medium Altitude Long Endurance) UAV is characterized by

Altitude:	medium
Payload typical:	200 kg
Endurance:	up to 30 hrs

Such a UAV is operated in the controlled airspace and must, therefore, be implemented in a full ATM/ATC environment. Since the UAV is operated above 1000 m is, i.e. in general, not below clouds, a radar (SAR) sensor is required, which could be complemented by an optical/IR sensor system. This, in turn, requires an appropriate payload capacity. From the technical system and platform point of view, all components for UAV based pipeline inspection are available:

- The EAGLE UAV has demonstrated successful operation, e.g. in campaigns in Northern Scandinavia in 2002;
- One possible (and tested) payload was a radar/SAR system, thus proving the respective feasibility;
- Due to the high radar data rates in the Gbit/s range the data transmission to ground stations requires an advanced IT technology, however this problem has been proven to be solvable;
- The image processing and feature extraction efforts is much more complicated than for scenario 1, since;
  (i) radar data require more sophisticated processing steps; and
  (ii) this data, in general has to be combined/fused with data from other sensors.

As in scenario 1 the certification standards, safety, and ATM/ATC regulations still have to be developed, which is again a necessary condition for the operation of a UAV based pipeline monitoring system.

For both UAV cases, the technical feasibility of the total system and platform is only the first step: Once all (future) requirements and standards for a safe operation are met, the crucial aspect of cost-effectiveness can be considered, which has to take into account all cost categories including potentially high expenses for certification and operation.

4.2. General Approach of Applications

Today the UAVs can successfully use in the appropriate platforms for a remote sensing based inspection system. Appropriate small and medium size UAV have been developed, their operation is technically feasible in controlled as well as in uncontrolled airspace. The technology to be intended for UAV sensors as well as for data collection and information processing seems to be suitable for this task. At the same time some of important and significant issues for successful operation of UAVs have to solve before a UAV based pipeline inspection system can be employed:

1. The data collection and information processing has to be developed to an operational standard, the capability of near-real-time determination of threats with high probability and low false alarm rates is not yet existing;
2. A total operational system consisting of UAV platform, sensors, data collection and data processing and alarm detection has not yet been demonstrated and proven its feasibility in a complete mission;
3. Finally the certification and operation standards for a safe and efficient operation of UAVs in controlled as well as uncontrolled airspace might be the most crucial problem.

It is understandable that in order to take benefit of the digital image data stream to be supplied by the imaging sensors of a satellite, aircraft or UAV should be analyzed by automated techniques, rather than by pure visual inspection. In the meantime the features, which have changed along the pipeline track and are possibly interfering with the buried pipe, should be automatically detected – based on the supplied digital imagery - and supplied to an alarming system. Possibly an easy task for a skilled human analyst, software to mimic human visual understanding is currently not used for this task. Traditional image analysis software are either targeted to evaluate the numeric signal contained in the single image pixel or to find a specific target based on a fixed image pixel template by means of template to image matching [17, 18, 19].

It is necessary to consider that the human visual system does not look at pixels, but immediately grasps the content of an image by means of analyzing the "objects" contained in an image and taking their "spatial relations" into account. Moreover, the information about "objects" is weighted according to their reliability. The sum of all this information (including the background knowledge of the analyst), determines the interpretation of an image. A new algorithmic approach is going to use exactly this procedure to find potentially objects in digital imagery which can be dangerous for pipeline infrastructure.

This approach is implemented in the software eCognition (REF). Within eCognition digital raster images are "segmented" to generate a hierarchy of objects, i.e. homogenous areas (such as houses, streets, but also cars on streets) are treated as "objects" and no longer as set of individual pixels. All identified objects are annotated with their color, color statistics, size, shape and neighborhood relations to other objects.

It is necessary to identify what kind of objects would be of interest, "rule bases" specify the visual appearance of objects are able to incorporate other information (such as the position of the pipeline corridor supplied in a GIS structure) and apply certain reliability weights to the defined information (e.g. a "car" has a minimum and a maximum size). The reliability is introduced by means of fuzzy logic reasoning. The therewith defined rules can now be applied to the network of image objects, automatically generated by eCognition. The result is the identification of the image objects defined in the rule base. Due to the fuzzy logic implied, identified image objects are labeled with a reliability of extraction. Based on the quality of the input image data and the unambiguous definition of rule bases, objects can carry certain degrees of "danger" to be imposed on the pipeline system. At a certain "danger" level, these objects are subject to further – possibly human – intervention and inspection. Because eCognition identifies objects with a clear outline, outlined of objects can directly be vectorized and exported to Geographic Information Systems (GIS) for further management and analysis.

In order to provide an initial tests to classify objects with the pipeline corridor it is needed a high resolution space borne optical imagery. Due to the use of shape and neighborhood relations in object oriented image classification, the dependence of spectral reflection of objects (in parts even the use of radar instead of optical images) can be diminished. Rule sets are identified which classify a scenery as independent as possible from the sensor technology used and are only (in operational system automatically) updated on certain functions based on specific parameters of the sensor (including image geometry) and the scenery (e.g. the sun elevation and the corresponding casting of shadows plays an important role in the identification of objects).

Amongst the operational advantages of pipeline monitoring is the general knowledge of the scenery, due to the frequent observation of the pipelines. An operational image analysis system has therefore less to concentrate to identify the entire scenery as such but to address the differential changes in the scenery which might have occurred during the last observation. As a multitude of sensors and imaging geometries would be used by an operational system, the "normalization" of the general scene information to a standard GIS format would be appropriate. The question is than to extract the changes within the updated imagery with reference to a GIS (and not the changes of one recent optical image with a previous radar image of the same scene). Again the advantage of object oriented feature analysis is the concentration on object appearance and neighborhood parameters and therewith the wide independence from imaging scale, geometry and sensor technology.

5. Pipeline Monitoring Based on Integrated Satellite Data

One of the largest and fastest growing markets for high-resolution satellite imagery in digital mapping is the utility industry. Utility companies and semi government bodies are turning to high-resolution satellite imagery to identify optimum facility and infrastructure locations. Using highly accurate, digital, orthorectified images, utility companies have a valuable information resource for planning, implementing and maintaining facilities and infrastructure, supporting disaster management efforts (Lindsay, 2001). High-resolution satellite imagery can aid utilities in monitoring electric and gas transmission corridors and rights-of-way; accurate and economical corridor analysis for power and gas pipelines; monitoring vegetation intrusion on transmission and distribution lines; and mapping for cable placement.

The integration of high-resolution imagery with geographic information systems to allow accurate geopositioning of pipeline and power line vector information on to the local land use and topography representation becomes a very useful planning and decision support tool. The pipeline location can be placed as a vector file over a one-meter spatial resolution satellite image and colored red. Sensitive environmental areas are then identified as green through a land classification analysis on the GIS product. The location of roads, agricultural areas and structures are also clearly distinguishable due to the high spatial resolution of the image.

Unauthorized intrusion by individuals onto power line or pipeline rights of way and the conduct of damaging activities to the pipeline infrastructure is a concern that has grown particularly in resent times. In North America there are thousands of miles of pipelines that are located in remote rural and wilderness areas. Monitoring of these vast networks on even an intermittent basis is difficult and costly. The most common monitoring method is to use small aircraft to fly over the rights of way on a monthly or less frequent basis and conduct a quick visual inspection. These methods provide a very limited monitoring effectiveness and even less security protection.

With the recent availability of high-resolution commercial satellite imagery, advanced radar system, and target detection algorithms significantly improved monitoring of power distribution networks are now possible. These imagery collects were sequenced in time and change detection analysis was conducted to identify potential security and encroachment events. In this example additional one-meter spatial resolution imagery was collected over the target area. In Figure 8. the three different imagery products are shown as overlays in the area of highest interest.

An example of this kind of imagery coverage and data analysis system for a pipeline network in is shown in Figure 9. In this application multi-spectral four-meter spatial resolution imagery was collected over larger areas of the pipeline system. In addition synthetic aperture radar (SAR) imagery was collected over the same area, which provides all weather, day and night coverage that is not possible with optical sensors. Multiple satellite systems can be tasked to image entire pipeline distribution networks on a daily basis. Time sequenced image analysis is then conducted using computerized change detection analysis to identify potential unauthorized encroachment, environmental risks and security problems. As shown in step four of Figure 9, before and after changes in imagery are identified and analyzed. Detected target information is reviewed and geo-referenced to map locations. Additional higher resolution imagery maybe collected over the geo-referenced locations if needed to determine if the situation calls for a notice, alert or alarm category of response. If determined necessary field personnel would be notified and dispatched to the predetermined location.

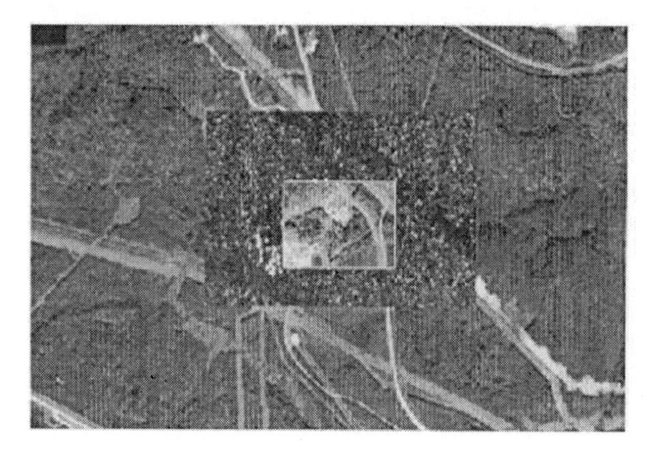

Figure 8. Multi-Sensor data fusion example fro pipeline monitoring

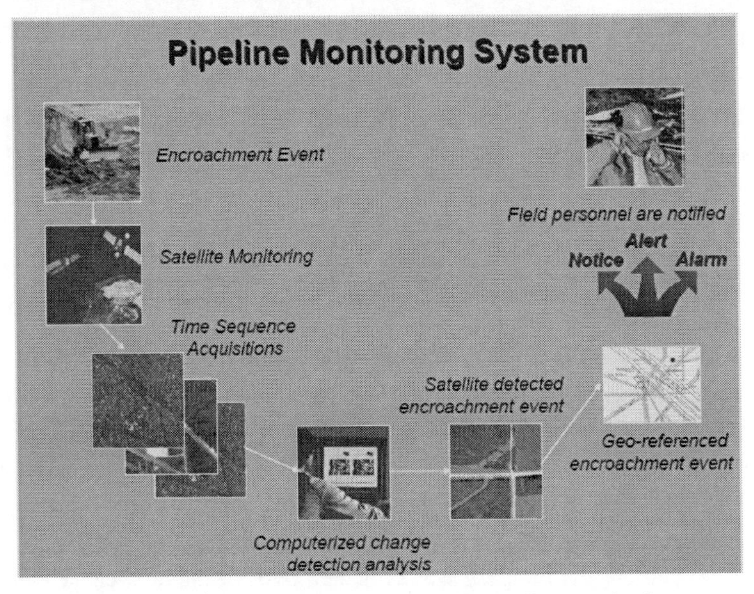

Figure 9. Satellite imagery based decision support system for pipeline monitoring

5.1. Effectiveness of Satellite Based Pipeline Monitoring System

Conventional monitoring of the massive pipeline systems in Canada is most commonly done on a sporadic schedule by used on visual observation from aircraft flying the lines. Typical flight observation coverage is in the order of once every one to two months for most of the pipeline miles. Detection of unauthorized intrusion and security risk events with current practice is low due to a number of limitations that are inherent with this approach to monitoring. These limitations include; infrequent coverage, human error in detection, limited data analysis capability, high mobilization costs and limited trend analysis.

In an effort to evaluate alternatives to airborne remote sensing of the pipelines a study was conducted to compare this mode of data collection with satellite based remote sensing. A summary comparison of the results is given in Table 1. Based on satellite coverage schedules three imagery collection frequencies were used. These included collection frequencies of once per week, twice per week and once per day. For each category the use of satellite based remote sensing was able to detect pipeline security problems 30 % to 100% better than aircraft based remote sensing. The most significant improvement was with once per week imagery collection where airborne systems had a 20% detection rate compared to a 32% to 55% detection rate for satellite systems. The highest rate of detection (93%) was also achieved with satellite systems with once per day imagery collection. For airborne systems the best detection rate of 88% was achieved using a twice per day imagery collection rate.

There are also challenges that may slow or impede the application of geospatial technologies to the electric utility sector. These include the need for improved methods and authorities for better data sharing across institutional boundaries. The developers and user communities need to communicate better and overcome some significant disciplinary differences. There are also challenging technical issues in the multi-sensor data fusion area to be overcome. Finally, there is a need for a focused interdisciplinary effort to match geospatial capabilities with specific user requirements.

Table 1. Comparison of aerial and satellite monitoring system for pipeline security applications

Frequency of Imagery Collection	Probability of Detection (%) with Aerial Sensor Systems	Probability of Detection (%) with Satellite Sensor Systems
Once per year	0.4%	
Once per 6 months	1.0%	
Once per 3 months	2.0%	
Once per month	5.0%	
Once per week	20.0%	32% to 55
Twice per week	40.0%	50% to 70%
Once per day	70.0%	78% to 93%
Twice per day	88.0%	

There are also challenges that may slow or impede the application of geospatial technologies to the electric utility sector. These include the need for improved methods and authorities for better data sharing across institutional boundaries. The developers and user communities need to communicate better and overcome some significant disciplinary differences. There are also challenging technical issues in the multi-sensor data fusion area to be overcome. Finally, there is a need for a focused interdisciplinary effort to match geospatial capabilities with specific user requirements.

Presently very sharply there is a question a safety of transportation of oil and gas in region of Azerbaijan/Georgia/Turkey after successful construction of an oil pipeline to the Baku- Tbilisi-Ceyhan and a gas pipeline Baku - Tbilisi - Erzurum. Construction of the railway of Baku - Tbilisi - Akhalkalaki - Kars has received the legal status after signing an Agreement between country representatives of the foregoing railway. Application space technology would become the most effective instrument for realization of safe functioning of this infrastructure [20, 21].

Each of these countries could bring the contribution for the positive decision of this problem. Certainly, the key position of Turkey which has an enough capacity and potential in the area of space science and technology could play a main role in this matter. Scientific and highly skilled experts of Azerbaijan and Georgia could provide appropriate benefit for realizations of the project.

5.2. Remote Sensing Methods and GIS Technology in Monitoring of the Linear Infrastructure for Environmental and Social Impacts

In accordance with this the Western Route Export Pipeline (WREP) used for oil transportation in Caucasus region was taken for application of remote sensing methods and space data processing. For this reason more sensitive areas of the WREP pipeline have been selected for investigations. The data has been bound to the space image throughout the topographical map. The final result of the work has reflected on the base of developed GIS.

In this part is intended to be building the stations for monitoring for the more sensitive areas among the pipeline. Within this approach an environmental issues can be successfully implemented which is a part of the safety and security aspects of the linear infrastructures. There is an availability of the combine the filed data as well as an aerial processed data which play the significant impact in pipeline problem solving.

The outcome of research is the mainly development of GIS risk analysis map of pipeline based on a filed measurements as the first step development of the pipeline network map of Azerbaijan Republic. Due to the existed oil and gas transportations pipeline systems it is undertaken not only produce a GIS pipeline map for the WREP as well as for the whole presently exploited infrastructure in the area.

The WREP corridor has the highest population density and the most concentrated transportation network. The lack of sufficient information among the pipelines can be a reason of disastrous consequences if an accident occurs (Figure 10).

The objective of the research has been undertaken for identification of the most sensitive zones among the pipelines corridor and based on the collected data development of Geographic Information System (GIS). Sensitive zones have been discovered using the field observation guided by existing pipeline information and application of Global Positioning System (GPS) technology. The spatially-referenced database is satisfied the need for accurate and easily accessible pipeline information. Field crew has been collected GPS point data at the pipeline crossing of public roads, railways and rivers within the studied sensitivity area. The GPS gathered data during the fieldwork were entered into the GIS. The GIS was used to generate statistical, spatial and thematic data. A comprehensive set of spatially-referenced pipeline risk zones information has been generated.

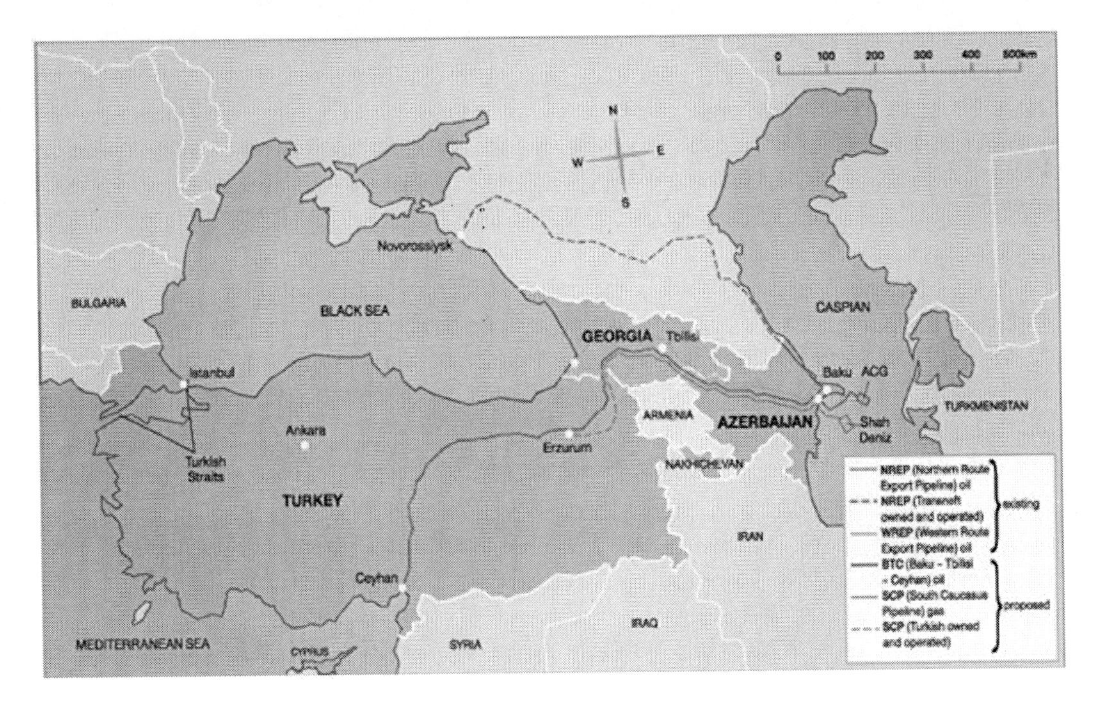

Figure 10. Western Route Export Pipeline (WREP) map of Azerbaijan

The steps of research achievements are following:

- Research – all pertinent and relevant existing data on the pipelines in the study area were collected and reviewed;
- Field Investigation – field crew using GPS units, collected latitude and longitude readings at the pipeline crossing of public roads, railways and rivers within the study area;
- GIS Development – GPS data collected by the field crew has been used fir development of pipeline system layers based on GIS;
- Outcomes – GIS has been developed with reflection of generation of the statistical, spatial, and thematic data. In the meantime hard-copy maps and digital GIS files were produced. The final product of the investigations was the unique database specifications as the digital base maps layers. This was done to ensure a seamless integration of the GIS pipeline map into the larger pipeline database.

The resulting spatially-referenced database should satisfy the need for accurate and easily accessible pipeline information for WREP. The database can be important tool for the protection and utilization of the significant resources.

Due to the existed maps of pipeline with inaccurate and/or too generalized information there is a need to develop the GIS for Azerbaijan. There is also a need for uniformity, since the maps contain varying levels of accuracy and scale. GIS pipeline mapping has many advantages over viewing spatial data on a map. GIS mapping allows for a level of analysis such as establishing networks creating buffers and modeling distance decay, which is not possible to produce on the hard copy maps [22, 23, 24].

It is a highly important to have a sufficient information related to the pipeline's location for the reason of a leak or rupture. In case of any pipeline leak whether large or small can be dangerous impact to the population in the area. This event can range from volatile components exploding to a large scale hazards. The pipeline systems cross many areas of alluvium along major rivers, streams, and bayous. These areas would be highly susceptible to contamination by liquid pollutants.

The need for development of the pipeline GIS for WREP is also related for population protection. On the studied within the 5 km buffer zone along the WREP are located urban settlements with population around 405 600 (Azerbaijan States Census Bureau 1999).

Knowledge of the precise location of these pipeline systems, sensitive zones is the significant issue for a many entities, including city and region authorities, emergency officials and business and industry representatives. This study provides the relevant and updated information being to satisfied needs of customers responsible for safely oil and gas transportation.

Methodology

At the first stage of investigation it was necessary to collect of the existing information regarding the pipeline systems for further understanding of the status. This information has been undertaken for GIS development for the WREP pipeline system.

In the next stage the maps were used for identification of the existing pipeline configurations in the investigated area. Based on this the present status of the pipeline systems and field investigation protocol were developed.

Field methodology

The field components of the study as the presence of pipelines, pipeline crossing with roads, railway, rivers as well as pump stations have been involved. Although pipeline sometime is visible it is usually only the presence of a "witness post" or a "right-of-way" that provides evidence for the field staff. Pipelines often follow "rights-of-way," long avenues of clearing where no development occurs. These rights-of-way are clearly visible on aerial photography. Pipelines are required to be marked by "witness posts" whenever they cross a road. For researchers studying pipelines, the witness post contains same useful information as the operator name and an emergency contact phone number.

Field staff was experienced in the use of GPS and GIS systems were employed to collect pipeline data throughout study area. These field crews were directed to collect GPS point data at the pipeline crossing of public roads, rivers and railway within the study area.

GIS development

The further stage after the GPS data, digital photographs, and other documentation were collected based on each field investigations and transferred into Microsoft Excel spreadsheet, a method was applied for data projection and conversion and subsequent GIS development.

Data projection and conversion

The deliverable of the digital data for this project must be seamlessly integrated with the existing data. This required that all digital data generated for this project be based on a Gauss Kruger coordinate system, Zone 8 using Pulkovo 1942 as the reference datum. Therefore, all data brought into the project GIS were converted to Gauss Kruger.

ESRI's ArcGIS/ ArcInfo 9 suite of GIS software was used throughout the work. The resulting deliverables in projects are in the form of ESRI .shp files. ArcGIS/ArcInfo is vector-based software that is the industry-standard for GIS-based research. ArcGIS software includes the following suite of integrated applications: ArcMap, ArcCatalog, and ArcToolbox as well as 3D Analyst, Spatial Analyst, Network Analyst, Geostatistical Analyst and needed for work other extensions. Each of these applications was valuable for conducting the GIS research for this work. ArcMap, a map-centric application was used for basic mapping and editing tasks, as well as analysis. ArcCatalog was used to manage geographic and tabular data and to create and edit metadata. ArcToolbox was used for data conversion and geoprocessing. ArcInfo, an enhancement to the standard ArcGIS software was also used because it provides advanced editing capabilities [25, 26].

A base GIS was developed using the following data sources:

shp files of digital vector data;
1 : 100 000 scale topographic maps (PULKOVO 1942 datum);
Mosaicked 12 LANDSAT Enhanced Thematic Mapper Plus (ETM+) scenes;

Digital orthophoto generated from QuickBird or IKONOS imagery for most sensitive zones in the study area.

Digital elevation model

Vector data consist of regions boundaries, urban settlements, roads, railroads, hydrologic features such as rivers, water reservoirs and lakes, high voltage electric power transmission lines and WREP pipeline data were selected because they are the most accessible, accurate, and detailed collection of digital data for study area as shown in Figure 11 and 12.

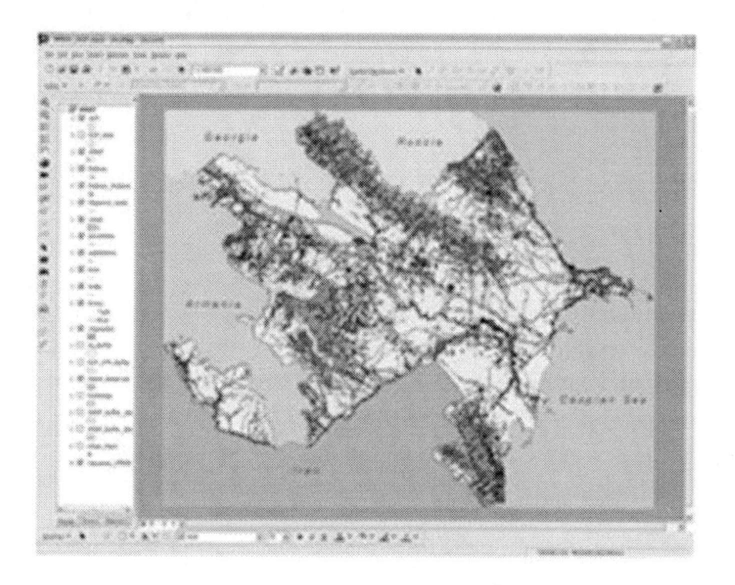

Figure 11. Geographic information systems project database in ArcMap

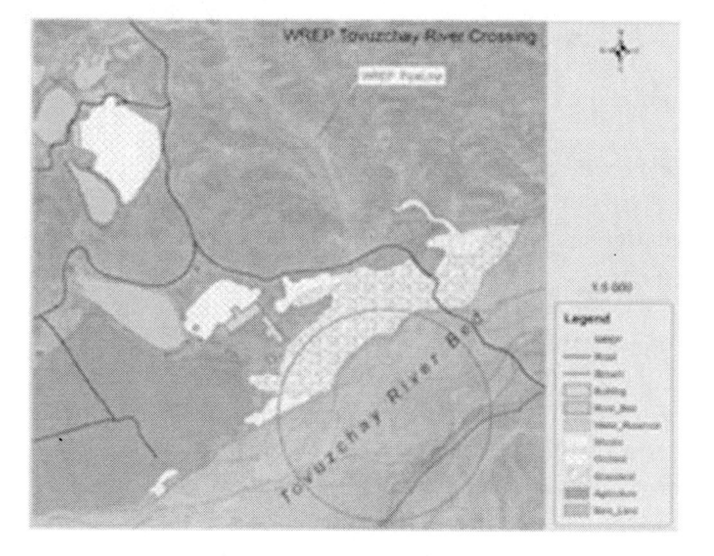

Figure 12. Land use around WREP - Tovuzchay River crossing

Figure 13. The orthophoto of Sangachal Terminal creted from QuickBird image

For more detailed analysis the sensitive zones around WREP pipeline are generated the orthophotos from very high spatial resolution satellite images like IKONOS (1 m) and QuickBird with spatial resolution 0.60 m (Figure 13). Hereafter for this purpose planed involve aerophoto imagery.

The coordinate values in the latitude and longitude columns of the .dbf file of GPS points representing the location of pipeline crossing and Pump Stations were displayed as x and y values using ArcMap. These values were exported as a .shp file and their projection defined as latitude/longitude (PULKOVO 1942) using ArcToolbox. Although these GPS points were stored in the latitude/ longitude reference system, the file displays properly when added to the project GIS. Because it is projected "on-the-fly" to Zone 8, PULKOVO 1942.

In the stage of the final editing process was completed and the vector layers in the .shp files representing different data had to be further processed in order to create topology. Topology is the procedure for defining spatial relationships, such as line connectivity, in a database.

A digital elevation model (DEM) is one of the most important datasets for analysis, management and planning. It provides elevation information that is useful for many environmental applications including landslide stability, hydrologic modeling and flood management planning.

The digital elevation model (DEM) was generated from digitized contour lines and elevation points from a 1:100 000 topographic maps.

It has a horizontal resolution of 50 m and a vertical accuracy of about 20 m. The accuracy of these DEMs may not be suitable for all application areas as it reflected in Figure 14. Therefore, accuracy assessment of current DEM was deemed necessary during field work. In this study existing ground survey marks were used as 'true' elevation data for DEM accuracy assessment. It is planed generate more high resolution and accuracy DEM for critical areas.

The three dimensional model of studied area generated in the ESRI's ArcScene application using digital elevation data and LANDSAT ETM+ sensor imagery. In this model the data can be viewed from different orientations and in a virtual fly-thru mode. The fly-thru capability allows the user to experience the terrain in a way that provides unique information about the terrain and a better understanding of the conditions on the ground (Figure 15).

In Figures 16 and 17 have been shown the results of development of road and river crossings of the WREP pipeline. Those crossings were based on the integration of the QuickBird imagies with the field data as one of the more sensitive areas of the pipeline itinearary.

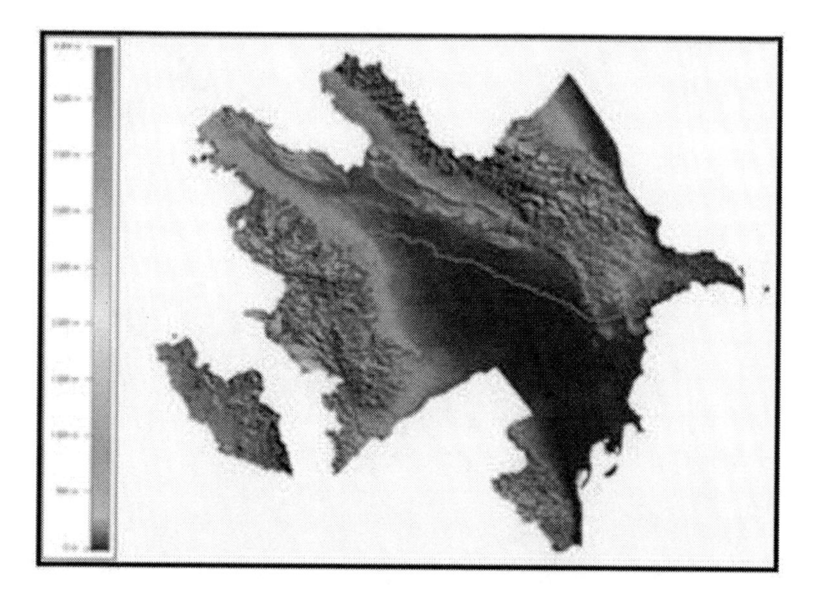

Figure 14. Digital Elevation Model (DEM) of Azerbaijan territory

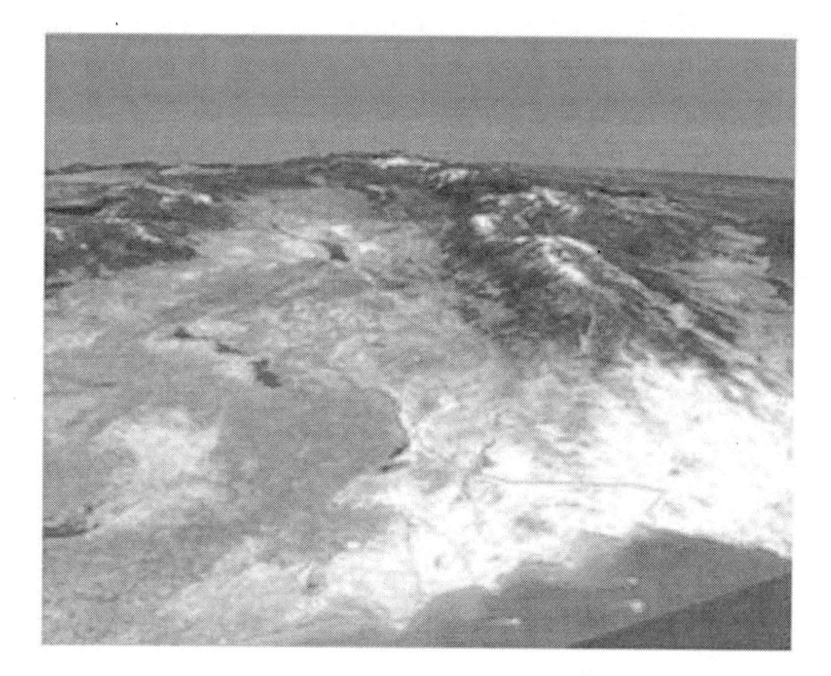

Figure 15. The 3D model of the WREP pipeline corridor

Figure 16. The QuickBird image of "Pump Station No 5"

Figure 17. The QuickBird image of "East Kura" river cross

6. Space Technology Applications in the Linear Infrastructure Security and Safety Purposes

The Company Gazcom has already announced that it is undertaken to develop a system for use of remote sensing methods for Earth observation. Within the framework of this project which is being privately funded and is expected to be paid by itself, intended cost from 300 to 400 million dollars as it was announced by Gaskom chief Nikolai Sevastyanov. "The company expects to invest 30% of the required fund and to borrow the other 70%". The new system which is being developed is to be used primary to conduct all weather damage monitoring of oil pipelines, to control land use, to prospect for a new oil fields, monitor emergency situations and assess environmental damage caused by industrial accidents and natural disasters. The launch of the first two space vehicles was scheduled for 2007. As more satellites are built the services will be offered to other large corporations and state bodies. The system is expected to have covered its outlays in three to five years (The Time Is Coming for Commercial Space Exploration, Russian Information Agency, April 13, 2005).

Definiens Imaging GmbH, leading supplier of object oriented Image Analysis Technologies announces, that its innovative eCognition technology was selected for feature recognition within the European Commission Pipeline REmote SENsing for Safety and the Environment (PRESENSE) project. A 4 million Euro, remote sensing, pipeline monitoring project, headed by Advantica Technologies Ltd and the European Gas Research Group GERG with 17 collaborating partners, has gained approval for funding from the European Commission 5th Framework Research and Technology Development programme, FP5.

PRESENSE aims to further improve the safe and secure transmission of gas in Europe's extensive high pressure gas mains transmission systems, by assessing the potential benefits of remote monitoring techniques and processes. The project will involve the development of mathematical modeling software, image processing software and the evaluation of detector instrumentation. A satellite remote monitoring system will beam images to Earth from satellites in space, mapping the route of a pipeline and identifying key threats to pipeline safety and security. Definiens Imaging will coördinate the "feature recognition" work package in this project and will use its innovative object oriented eCognition technology to automatically find potential targets which might compromise pipeline security. eCognition has been selected due to its unparalleled capabilities to fuse vector and raster information, to identify objects and to put these objects into a spatial context. It is expected to provide eCognition to the PRESENSE Consortium and being selected as a key technology provider in this project. In the meantime beyond the standard environmental applications of Earth observation, the new series of high resolution satellites open a new opportunities for monitoring facilities which can be supported by eCognition.

The combination of object identification and a semantic knowledge network appeared to be in image processing procedure which is especially well suited for pipeline monitoring. The commercial analysis system eCognition allows data from different sensors (optical and radar) to be merged and combined with geographic information system data for object identification [27]. The eCognition method includes the identification and generation of objects from the original pixel-based files and establishment of semantic links between these objects and known features, for example in the form of a feature database. Image features such as vehicles or pits are classified on the bases of radiometric, geometric and other links between the image objects and placed in relation to neighboring objects and known information from geographic information systems. Any vehicles detected not on defined roads but in open country near to the coordinates of a pipeline route could therefore be identified as potential hazards. By assessing the area concerned, assumptions can be made concerning a possible distinction between agriculture vehicles and construction equipment.

Pixel-based image processing techniques are often unable to recognize characteristics that are obvious to human visual inspection. This because pixel-based image classification uses the spectral information represented by the digital numbers in one or more spectral bands and attempts to classify each pixel based mainly on this spectral information. The eCognition's unique approach is based on a simple concept: important semantic information necessary to interpret an image is not represented in single pixels but in meaningful image objects and their mutual relations. The eCognition software allows the analysis of image objects instead of single pixels. A hierarchical image object network is used to select and combine appropriate scales for image classification. Based on this image object network neighborhood and context can be integrated. Furthermore fuzzy classification is a part of the system so that expert knowledge can be implemented and inherent uncertainties of remote sensing data can be taken into account.

The first step for optical change detection can be described as a post-classification change detection algorithm in which changes are detected based on the results of land cover classification. Changes occur if the land cover between two images is different. For that Basic Land Cover Classification optical data and a thematic GIS layer are imported in the software eCognition. Then image objects are created by a patented algorithm called Multiresolution Segmentation. These objects have certain attributes that are used in the next step for classification based on a hierarchical knowledge base. The knowledge base uses attributes and their combinations to describe different classes. In the optical imagery some of the small objects that had to be detected could only be identified because of the shadow they cause. Thus in the optical change detection approach elevation information is used together with optical data to distinguish between expected shadows and expected shadows. Only if a shadow is located in a certain position tp an elevated object dependent on the sun angle it is classified as an expected shadow. In case the object is not elevated according to the elevation information and therefore a new object that causes an unexpected shadow this object is labeled in the preliminary change detection as possible hazards. Additionally as a final step object size can be taken into account. Thus large and small machines can be distinguished [28, 29].

6.1. The Pipeline Information Management System (PIMS)

Demonstration of a change detection and alarm handling system was one of the key targets of the PRESENSE project. The objective was to show how an integrated and fully automated pipeline monitoring system could provide the level of detail required by the pipeline operators in response to real alarm scenarios, whilst minimizing the number of false alarms. The prototype information system comprises four linked components:

- Pipeline Operator System (POS);
- Pipeline Information Management System (PIMS);
- Hazard Extraction System (HES);
- Image Collection System (ICS)

The area of interest surrounding a pipeline is defined within the knowledge base of the Pipeline Information Management System (PIMS). PRESENSE will detect changes in features within this area using Earth Observation imagery and categorize these changes into priorities. Data layers from many sensor technologies as well as basic map data and GIS information are geo-referenced and stacked, increasing the reliability of the change detection process. Detected changes are then classified within the HES into potential hazards, with priority levels, by reference to land classification and the PIMS. These hazards are presented to the pipeline operator through a web browser front end of the Pipeline Operator System (POS). The operator can interrogate the system through the web browser or in more depth through a further Graphical User Interface (GUI). The operator can analyse the alarm in further detail and review its history and finally provide the system with information on how the alarm has been handled e.g. by sending a pipeline inspector to the site in question. The pipeline database is then updated to reflect the current situation.

6.2. Results of Presense

The PRESENSE system was successfully demonstrated in September 2004 using the previously acquired test site data. The consortium partners concluded that further developments are required on several fronts before a commercially viable system can be achieved:

- A need for better satellite resolution for both optical and SAR;
- A need for more satellites carrying either optical, SAR or both sensor technologies;
- Ideally satellite borne Lidar;
- Lower cost imagery available on-call;
- Reduction in false alarm rates through improved algorithms.

At the same time current pipeline monitoring is undertaken routinely using a number of techniques, the most common being "flying-the-line" in helicopters or fixed wing aircraft with trained observers on board. This happens with regular frequency but is expensive and has the inherent safety concerns of all helicopter flights. PRESENSE brings together several of the major European gas utilities and expertise from a number of the European space research agencies, together with organizations with specialist capabilities in information computing technologies such as image analysis, data management and dissemination. This provides a unique forum where new approaches to pipeline monitoring and the resulting benefits can be fully explored.

The two and a half year programme was commenced on the 1st January 2002. Several of the partners were provided state-of-the-art monitoring equipment that was flown over pipeline test sites to assess the performance of new a techniques and the processing of multi-sensor data sets. Results were compared with imagery collected from current satellite instrumentation to enable the technology limits of the various approaches to be compared. The pipeline operator companies were provided access to their pipeline right-of-ways and was developed models for the risk and cost benefits of the various approaches considered. An outcomes of the project were confident that it would be able to demonstrate the capabilities of remote monitoring from space and the huge benefits of the system for the European gas utilities as compared to current methods of network monitoring.

It is very important to indicate that eCognition software provides object oriented multi-scale satellite image analysis capabilities, and classifies an image based on attributes of the image objects rather than on the attributes of individual pixels. This classification process allows users to collect a large range of statistical parameters and detailed information from any Earth-observation data and any raster or vector based information tailored specifically to their information needs.

There is other opportunity of Definiens Imaging GmbH (www.definiens-imaging.com). Definiens Imaging GmbH was established in 2001 and incorporates all Geomatics and

Imaging technologies and products of Definiens AG (www.definiens.com). Definiens AG evolved in September 2000 from Delphi2 Creative Technologies GmbH. The company was founded in 1994 by Nobel Laureate Professor Gerd Binnig and science journalist Dieter Herold. Definiens AG develops decentralized communication platforms as well as software for knowledge management and image analysis. Definiens AG's solutions are based on the "Cognition Network" model. The "Cognition Network" uses technology that closely emulates human thought patterns. Definiens AG and Definiens Imaging GmbH are based in Munich, Germany. eCognition was awarded the European Information Society Technology prize in 2001.

The Advantica (www.advanticatech.com) is a part of the Lattice Group – the infrastructure technology group which also includes Transco. With origins in British Gas, Advantica is a leading provider of technology and engineering services to customers in gas, pipelines and associated industries internationally. Advantica recently made two acquisitions:

- RISX – the Aberdeen-based risk management consultancy; and
- US headquartered.

Stoner Associates – the number one network simulation software company in the US. Stoner specializes in providing off-the-shelf products for the gas, oil, petroleum and water industries primarily in the US where two thirds of its approximate 550 clients are located. Stoner software is also used by clients in 48 countries outside North America. With Stoner's excellent products and extensive client base, the acquisition will make a significant contribution to Advantica's future success in the US. Advantica now employs 1,000 people worldwide – predominantly in the UK and North America. The company's largest office is located in Loughborough, UK. Advantica's turnover is predicted to be £94million this year.

The consortium, managed by Advantica Technologies Ltd, on behalf of Transco – operator of the gas network in Great Britain - comprises the following partners:

- Seven further European gas & oil pipeline operators, including BP (UK);
- Distrigas -Now Fluxys NV-(BE) Gasunie (NL);
- Gaz de France (F);
- Ruhrgas and VNG (D).

The consortium manages extensive pipeline infrastructures within Europe and beyond. Four "Remote Sensing Technology Developers", mainly from the aerospace and remote monitoring industries of Europe, comprising NERC (BGS) from the UK, CS from France, DLR from Germany and TNO from the Netherlands, that will provide expertise in the areas of surveillance platforms & equipment to provide primary data on recognition and analysis of known features on the pipeline. The University of Nottingham will provide vegetation stress analysis activities. Two further providers of remote sensing technologies and services, AES from Germany and NPA from the UK. A further grouping of three technology infrastructure providers, bringing further IT support for the project, comprising Definiens Imaging from Germany, ISS from the UK and NLR from the Netherlands [30].

It is necessary to indicate a significant attempt of NASA led's technology and process integration efforts resulting in following areas:

- Enhanced UAS platform utility;
- Development and integration of new requirements- driven sensor technologies;
- Maturing real-time data telemetry and autonomous real-time processing; and
- Innovative data visualization and evaluation capabilities.

Stakeholders from the wildfire management and science community, the Tactical Fire Remote Sensing Advisory Committee (TFRSAC), guide development efforts by detailing data and information gaps, defining priorities for wildfire management capability enhancements, and smoothing the infusion of new technologies and processes into operational practices. The Western States Fire Mission (WSFM) demonstrated key elements of the WRAP project, showcasing an end-to-end solution set for providing real-time disaster-related information to Wildfire Incident Management Teams, including:

- The first time integration of a large UAS into the National Airspace for long-endurance/range data gathering;
- The development and transfer of a disaster-specific imaging sensor, tailored to improved wildfire observations;
- The development and demonstration of a data management decision support system derived from a NASA Exploration Program software capability, enhanced using a COTS visualization package for ease-of-use by disaster teams, and;
- The development and integration of software technologies for on-board autonomous processing of sensor-derived imagery and real-time telemetry of such to disaster decision makers.

The highlights of the WRAP and WSFM during 2007 were included:

- First-ever flights of a UAV in the NAS over extended regions of the continental western US for disaster support;
- UAS / sensor collections over 60 fire events on eight missions with real-time data delivered to each disaster management team;
- Flew four UAV missions of “emergency support” to the Southern California wildfires of October 2007, integrating seamlessly and flawlessly into the busiest National Airspace (NAS) in the country;
- Provided real-time data to eleven (11) ICC teams on the major SoCA wildfires, as well as to three county-level Emergency Operations centers (EOC)(San Diego, Orange, Riverside counties);
- During the eight missions, the provision of real-time data to ICCs saved million of dollars in personnel time, fire-fighting costs, property, and natural resource loss;
- Demonstrated robustness of UAV, sensor, and processing system during eight missions of 8-21 hours duration with over 2,000 data sets collected and delivered with no data loss or system failures;
- Capabilities demonstrated are in adaptation by partners for operational utility.

NASA Ames Research Center and BP Pipelines and Logistics North America conducted a preliminary exploration and testing/demonstration project in 2007 that:

- Conceptualized a longer-term effort to more efficiently monitor pipeline rights-of-way and improve the detection of intrusions and leaks;
- Tested/demonstrated potential methane detection and camera-based surveillance systems.

The SENTRI-LD Concept project undertakes dubbed "SENTRI-LD", calls for the development and demonstration of the systems and operational processes for remotely detecting intrusions into pipeline rights-of-way and leaks from liquid/gas pipelines via sensors and imaging systems on small manned and unmanned aircraft.

Intrusion Detection Camera Evaluation Study - NASA and BP teamed to evaluate performance of several camera systems for suitability for automated intrusion detection in realistic environment:

- To provide a basis for determining if Mercury Computer Systems' (MCS) cameras can provide images that allow a ground-based human operator to identify threats;
- To provide a basis for determining if MCS cameras can provide images of acceptable quality for the RoW archival video record;
- To assess performance of MCS anomaly detection software, vs. requirements;
- To provide a basis for determining if NASA-provided cameras can provide images that allow a ground-based human operator to identify threats;
- To provide a basis for determining if NASA-provided cameras can provide images of acceptable quality for the RoW archival video record;
- To provide a basis for determining acceptable spatial resolution for ROW monitoring by a human image interpreter;
- To collect imagery to support development of NASA detection algorithms.

CONCLUSION

Maintenance the security and safety aspects of critical infrastructures of power supply particularly in pipelines for transportations of oil and gas is a problem of rather actual. This problem has the significant place in energy as an important element of oil and gas transportation. For this reason Russian Federation as the main supplier of oil and gas is considering of maintenance of energy resources to the costumers safely. European Union and United States of America are undertaking to apply an advance technology for this issue. Taking into account significance of the problem leading countries have developed appropriate regulations and legislations for security and safety aspects of energy resources transportations.

Undoubtedly the use and application of space technology as the best instrument for data collection with a further geographically information integration is an excellent opportunity for related to this issue problem solving. This data is collecting by means of remote sensing methods based on Earth surveillance. There are requirements needed to be undertaken in data receiving and processing for the purpose of the safety and security issues of oil and gas transportation.

Application of advances in space technology demands to develop a new approach of detection technology and development a new detector systems meeting the main Earth observation requirements of linear infrastructure. This technology with an obvious advantage successfully can be found out an application not only for the general monitoring of the linear systems as well as a more sensitivity zones and areas of the infrastructures. It can be used for landslide and land movement identification and classification purposes.

It is considered in this chapter of use of the unmanned aerial vehicles (UAV) for the safely transportation of energy sources to costumers. The main target of this approach is to describe of necessity of application of such systems for the observation of linear infrastructures and demonstrate of advances and weaknesses of this technology.

As an example of maintenance of security and safety issues of the linear systems of oil and gas transportation is provided for the Western Route Export Pipeline. It has been developed a data based on the remote sensing methods and processed data was integrated to the GIS. It can be successfully undertaken for human and social aspects in energy sources transportation.

This chapter is considered to be useful for the appropriate officials for decision making. It is a source to collect a necessary high and accurate data for management within the state level in case of disaster and risk evaluation and assessment.

REFERENCES

[1] European Commission, *White Paper Space: a new European Frontier for an expanding Union*, COM(2003)673, Brussels, November 2003.

[2] European Communities; Research for a Secure Europe: *Report of the Group of Personalities in the Field of Security Research*, ISBN 92-894-6611-1, Brussels, 2004.

[3] Dekker, RJ; Lingenfelder, I; Brozek, B; Benz, UC; Van den Broek, AC. Object-based detection of hazards to the European gas pipeline network using SAR images. *Proceedings of the 5th European Conference on Synthetic Aperture Radar (EUSAR)*, Ulm, Germany, 25-27 May 2004.

[4] Schreiera, G; Hausamanna, D; Lingenfelderb, I; Benzb, U; Zirnigc, W. *a German Aerospace Center*, DLR, 82234 Oberpfaffenhofen, Germany.

[5] European Communities; Research for a Secure Europe: *Report of the Group of Personalities in the Field of Security Research*, ISBN 92-894-6611-1, Brussels, 2004.

[6] Zirnig, W; Hausamann, D; Schreier, G. A concept for natural gas transmission pipeline monitoring based on new high-resolution remote sensing technologies. *Proceedings International Gas Research Conference* IGRC 2001, Amsterdam, The Netherlands, November 2001.

[7] Remote Sensing and GIS Applications for Pipeline Security Assessment, William E. Roper and Subijoy Dutta, http:// proceedings. esri. com/ library/ userconf/ proc05/papers/pap1762.pdf.

[8] The PRESENSE and PIPEMON projects – defining the ways of using space-borne earth observation services for pipeline monitoring, E. ON Ruhrgas Aktiengesellschaft, Germany, Russ Pride, Advantica Ltd, United Kingdom, Iris Lingenfelder, Definiens

Imaging GmbH, Germany, Richard Chiles, Nigel *Press Associates,* United Kingdom, Dieter Hausamann, *German Aerospace Centre DLR*, Germany.

[9] Remote Sensing and GIS Applications for Pipeline Security Assessment William, E. Roper and Subijoy Dutta, http:// proceedings. esri. com/ library/ userconf/ proc05/ papers/pap1762.pdf

[10] Jeljeli, MN; Russell, JS; Meyer, HWG; Vonderohe, AP. "Potential applications of geographic information systems to construction industry." *Journal Construction Engineering and Management*, Vol. 119, no.1, 72-86, March 1993.

[11] Chang, KT. *Introduction to Geographic Information Systems*, Tata McGraw-Hill, New Delhi, 2002.

[12] Hausamann, D; Zirnig, W; Schreier, G; Strobl, P. Monitoring of Gas Pipelines – A Civil UAV Application, *Aircraft Engineering and Aerospace Technology*.

[13] Hausamann, D. "Civil Applications of UAVs – User Approach", *Shephard's Civil UAV Symposium*, London, UK, 17 – 19 July (2002).

[14] Hausamann, D; Brokx, W. "User Driven UAV Applications - Pipeline Monitoring and other Examples", *Proc. First European Conference on the Applied Scientific Use of UAV Systems*, Kiruna, SW, 10 - 11 June (2002).

[15] Zirnig, W; Hausamann, D; Schreier, G. "A Concept For Natural Gas Transmission Pipeline Monitoring Based on New High-Resolution Remote Sensing Technologies", *Contributed Paper, International Gas Research Conference 2001*, Amsterdam 5th – 8th November, 2001.

[16] Zirnig, W; Hausamann, D; Schreier, G. "High-Resolution Remote Sensing Used to Monitor Natural Gas Pipe-lines", *Earth Observation Magazine*, 2002, 11, 12-17.

[17] eCognition, 2003. most recent information under www.definiens-imaging.com.

[18] Benz, UC; Schreier, G. OSCAR - object oriented segmentation and classification of advanced radar allows automated information extraction. In: *Proceedings of IGARSS* 2001, July 2001, Sydney.

[19] Blaschke, T; Strobl, J. What's wrong with pixels? Some recent developments interfacing remote sensing and GIS. In: GeoBIT/GIS, 2001, 6, 12-17. http://www.definiens-imaging.com/down/GIS200106012.pdf.

[20] Sefer Kurnaz, Rustam B. Rustamov Regional Earth Observation Remote Sensing and GIS Services for Monitoring of Integration Systems, *Proceedings of 3rd International Conference on Recent Advances in Space Technologies, IEEE*, Istanbul, Turkey, June 14-16, 2007, 268-271.

[21] Sefer Kurnaz, Rustam, B. Rustamov, Maral Zeynalova and Fazila Qasimova. Natural growth as an indicator of monitoring of the linear infrastructure for safety and security issues. *KSAS International Journal (The Korean Society for Aeronautical & Space Sciences)*, May 2008, v.9 № 1, 59.

[22] Sefer Kurnaz, Rustam, B. Rustamov. Application of Space Technology in Support of Security and Safety of Critical Infrastructure, H; Gonca Coskun, et al. (eds.), *Integration of Information for Environmental Security*, 149-154, ©2008 Springer, The Netherlands.

[23] Rustam, B. Rustamov. History of Steps on Establishment Space Science and Technology in Azerbaijan. *KSAS International Journal (The Korean Society for Aeronautical & Space Sciences)*, May 2008, v.9 № 1, 48.

[24] Sefer Kurnaz, Rustam, B. Rustamov, Maral Zeynalova and Fazila Qasimova. Natural growth as an indicator of monitoring of the linear infrastructure for safety and security issues. *KSAS International Journal (The Korean Society for Aeronautical & Space Sciences)*, May 2008, v.9 № 1, 59.

[25] Maral, H; Zeynalova, Rustam, B. Rustamov, Saida E. Salahova, *"Space Technologies for the Benefit of Human Society and Earth"*, P. Olla, (ed.), DOI 10.1007/978-I-4020-9573-3_5, © Springer Science+Business Media B.V. 2009.

[26] Sefer Kurnaz, Gunel Babayeva, Rustam B. Rustamov, Elman Aleskerov, Monitoring of the Linear Infrastructure: Environmental and Social Impacts, *Proceedings of 4th International Conference on Recent Advances in Space Technologies, IEEE*, Istanbul, Turkey, June 11-13, 2009, 273-276.

[27] *Definiens Imaging GmbH: eCognition User Guide*. Munich, Germany, 2003.

[28] Van den Broek, AC; Smith, AJE; Dekker, RJ; Steeghs, TPH. Target Acquisition Performance as a Function of Resolution using Radar Change Detection. *Proceedings of the 5th European Conference on Synthetic Aperture Rada (EUSAR)*, Ulm, Germany, 25-27 May 2004.

[29] Dekker, RJ; Lingenfelder, I; Brozek, B; Benz, UC; Van den Broek, AC. Object-based detection of hazards to the European gas pipeline network using SAR images, *Proceedings of the 5th European Conference on Synthetic Aperture Radar (EUSAR)*, Ulm, Germany, 25-27 May 2004.

[30] *Definiens Imaging GmbH*, Trappentreustr. 1, 80339, Munich, Germany . pr@definiens.com, www.definiens-imaging.com, Munich – 04th January 2002.

In: Space Policy and Its Ramifications
Editor: John P. Ramos

ISBN: 978-1-61761-555-9

Chapter 2

THE FUTURE OF NASA: SPACE POLICY ISSUES FACING CONGRESS *

Daniel Morgan*

SUMMARY

For the past several years, the priorities of the National Aeronautics and Space Administration (NASA) have been governed by the Vision for Space Exploration. The Vision was announced by President Bush in January 2004 and endorsed by Congress in the 2005 and 2008 NASA authorization acts (P.L. 109-155 and P.L. 110-422). It directed NASA to focus its efforts on returning humans to the Moon by 2020 and some day sending them to Mars and "worlds beyond." The resulting efforts are now approaching major milestones, such as the end of the space shuttle program, design review decisions for the new spacecraft intended to replace the shuttle, and decisions about whether to extend the operation of the International Space Station. At the same time, concerns have grown about whether NASA can accomplish the planned program of human exploration of space without significant growth in its budget.

A high-level independent review of the future of human space flight, chaired by Norman R. Augustine, issued its final report in October 2009. It presented several options as alternatives to the Vision and concluded that for human exploration to continue "in any meaningful way," NASA would require an additional $3 billion per year above current plans.

In its FY2011 budget request, the Obama Administration proposed cancelling the Constellation spacecraft development program and eliminating the goal of returning humans to the Moon. NASA would instead rely on commercial providers to transport astronauts to Earth orbit, and its ultimate goal beyond Earth orbit would be human exploration of Mars, with missions to other destinations, such as visiting an asteroid in 2025, as intermediate goals. Operation of the International Space Station would be extended to at least 2020, and long-term technology development would receive increased emphasis.

---

* This is an edited, reformatted and augmented edition of a United States Congressional Research Service publication, Report R41016, dated April 19, 2010.

* dmorgan@crs.loc.gov, 7-5849

Committees in the House and Senate have held hearings to consider both the Augustine report and the Administration proposals. As Congress considers these broad space policy challenges, it faces choices about whether NASA's human exploration program is affordable and sufficiently safe,and if so, what destination or destinations it should explore; whether the space shuttle program should continue past its currently plannedtermination at the end of 2010 (or in early 2011); if so, how to ensure thecontinued safety of shuttle crews after 2010; if not, how the transition of theshuttle workforce and facilities should be managed; whether U.S. use of the International Space Station should continue past itscurrently planned termination at the end of 2015; whether the currently planned Orion crew capsule and Ares rockets, beingdeveloped as successors to the space shuttle, are the best choices for deliveringastronauts and cargo into space, or whether other proposed rockets or commercialservices should take their place; and how NASA's multiple objectives in human spaceflight, science, aeronautics, and education should be prioritized.

Introduction and Legislative Context

The idea of human spaceflight beyond Earth orbit has captivated many Americans for more than half a century. As U.S. space policy has evolved, new opportunities have emerged, and new challenges have arisen. For the past several years, the priorities of the National Aeronautics and Space Administration (NASA) have been governed by the Vision for Space Exploration. The Vision was announced by President Bush in January 2004 and endorsed by Congress in the 2005 and 2008 NASA authorization acts (P.L. 109-155 and P.L. 110-422). It directed NASA to focus its efforts on returning humans to the Moon by 2020 and some day sending them to Mars and "worlds beyond." The resulting efforts are now approaching major milestones, such as the end of the space shuttle program, design review decisions for the new spacecraft intended to replace the shuttle, and decisions about whether to extend the operation of the International Space Station. At the same time, concerns have grown about whether NASA can accomplish the planned program of human exploration of space without significant growth in its budget. When President Obama took office in January 2009, most analysts expected the new Administration to revisit the Vision, but it was not immediately clear what changes it would make.

In May 2009, when the Administration released the full details of its budget request for FY2010, it announced plans for a high-level independent review of the future of human space flight, chaired by Norman R.Augustine.[1] Major components of the NASA budget request were placeholders, to be revised following the results of this review. The Augustine committee released its final report in October 2009.[2] Committees in the House and Senate held hearings to consider the proposals.[3] TheAdministration did not submit a revised FY2010 budget for NASA. In December 2009, Congress appropriated FY2010 funds for NASA at

[1] For more details, see "Human Spaceflight: The Augustine Committee," below.

[2] Review of U.S. Human Spaceflight Plans Committee, *Seeking a Human Spaceflight Program Worthy of a Great Nation*, October 2009, online at http://www.nasa.gov/pdf/396093main_HSF_Cmte_FinalReport.pdf.

[3] House Committee on Science and Technology, *Options and Issues for NASA's Human Space Flight Program: Report of the "Review of U.S. Human Space Flight Plans" Committee*, hearing September 15, 2009; Senate Committee on Commerce, Science, and Transportation, Subcommittee on Science and Space, *Options from the Review of U.S. Human Space Flight Plans Committee*, hearing September 16, 2009.

approximately the level in the President's original request. The appropriations conference report (H.Rept. 111-366) stated that the Augustine committee's report

> raises issues requiring thoughtful consideration by the Administration and the Congress. It is premature for the conferees to advocate or initiate significant changes to the current program absent a *bona fide* proposal from the Administration and subsequent assessment, consideration and enactment by Congress. It is the expressed hope of the conferees that the Administration will formulate its formal decision soon, submit its recommendations for congressional review and consideration, and budget the necessary resources.

As part of its FY2011 budget, the Administration proposed major changes to the Vision, including the elimination of a human return to the Moon as NASA's primary goal, the cancellation of NASA's Constellation spacecraft development program, a new effort to encourage the private sector to develop commercial crew launch services, and increased emphasis at NASA on technology development and science. Congress will consider these proposals as it acts on FY2011 appropriations legislation.

The 111th Congress is also widely expected to consider a new NASA authorization bill. The 2008 authorization act authorizes appropriations for NASA only through FY2009 and still reflects congressional endorsement of an unmodified Vision for Space Exploration. The House Committee on Science and Technology has publicly announced its intent to "work with the new Administration on a multi-year authorization for NASA."[4] No new authorization bill has yet been introduced in either chamber, however.

As Congress considers these broad space policy challenges, the major issues it faces can be summarized as three broad questions:

- **What is NASA for?** Different analysts and policy makers give different answersto this question: making scientific discoveries, developing technologies witheconomic benefits, enhancing national security, enhancing international prestige,even fulfilling human destiny in space. How should these competing goals beprioritized?
- **What should NASA do?** In order to accomplish its broad goals, how shouldNASA balance its major programs in human spaceflight, robotic spaceflight,aeronautics research, and education? In the human spaceflight program, which islarger than all the others put together, should the agency's goal be exploration ofthe Moon, or some other destination? What should the top priorities be forNASA's science and aeronautics programs?
- **How?** Once these questions are decided, how should their answers beimplemented? What new space vehicles are needed? What should be done withexisting programs, such as the space shuttle and the International Space Station?

This chapter analyzes these questions and some possible answers. It also addresses a number of cross-cutting issues, such as NASA's interactions with other federal agencies and the growing role of the commercial space industry.

---

[4] House Committee on Science and Technology, *111th Congress – Agenda Overview*, http://democrats.science.house.gov/Media/File/ForReleases/111thSTAgenda.pdf.

What Is NASA for?

During the Eisenhower Administration, after the Soviet Union's launch of the first artificial satellite, Sputnik, but before the establishment of NASA, the President's Science and Advisory Committee identified four "principal reasons for undertaking a national space program":

- "the compelling urge of man to explore and to discover";
- "defense ... to be sure that space is not used to endanger our security ... [and to]be prepared to use space to defend ourselves";
- to "enhance the prestige of the United States ... and create added confidence inour scientific, technological, industrial, and military strength"; and
- "scientific observation and experiment which will add to our knowledge andunderstanding of the Earth, the solar system, and the universe."[5]

To these objectives, analysts today add

- the potential for technologies developed for the space program to have direct andindirect ("spinoff") economic benefits;
- the opportunity to use space activities as a tool of international relations, throughcollaboration on projects such as the International Space Station; and
- the ability of the space program to inspire students and promote education inscience, technology, engineering, and mathematics (STEM).

These goals form a foundation for U.S. space policies, but policy makers differ in how they should be balanced against each other. Is the urge to discover a sufficient reason to explore space, or must exploration also meet needs here on Earth? Should economic benefits be an explicit focus for NASA or just a positive side effect? To what extent should improving STEM education be a NASA function, as opposed to a consequence of its other functions? Should the emphasis of international space programs be competition or cooperation?

The priorities that Congress assigns to these objectives may determine how it balances the competing demands of NASA's programs. For example, if Congress believes that national prestige is a high priority, it could choose to emphasize NASA's high-profile human exploration activities, such as establishing a Moon base or exploring Mars. If scientific knowledge is a high priority, Congress could emphasize unmanned missions such as the Hubble telescope and the Mars rovers. If international relations are a high priority, Congress could encourage joint space activities with other nations. If economic benefits are of interest, Congress could focus on technological development, linking NASA programs to the needs of business and industry.

A report by the National Academies proposed goals similar to those listed above and recommended three criteria to use in balancing their competing demands for resources:[6]

---

[5] President's Science Advisory Committee, *Introduction to Outer Space*, March 26, 1958, http://www.hq.nasa.gov/office/pao/History/monograph10/doc6.pdf. For more information on the evolution of space policy since Sputnik, see CRS Report RL34263, *U.S. Civilian Space Policy Priorities: Reflections 50 Years After Sputnik*, by Deborah D. Stine.

- *Steady progress.* Each major area should be maintained at a level that allowssustained long-term progress with intermediate goals achieved at a reasonablepace.
- *Stability.* Rapid downsizing and abrupt redirection should be avoided becausethey are disruptive, can take time to recover from, and can create risk asoperations experience is lost.
- *Robustness.* Sufficient human resources and research infrastructure should bemaintained so that the nation can ramp up selected activities quickly in responseto changing national needs or scientific breakthroughs.

The Academies report did not, however, actually employ these criteria to prioritize the goals it proposed.

WHAT SHOULD NASA DO?

Based on this wide variety of objectives, NASA has established programs in human spaceflight, science, aeronautics, and education. The largest and most visible effort, in human spaceflight, faces considerable uncertainty about its proper scope and aims. The content of the science, aeronautics, and education programs is less controversial but still faces questions about scope, balance, and other issues.

Human Spaceflight: The Vision for Space Exploration

The Vision for Space Exploration, announced by President Bush in a speech on January 14, 2004, directed NASA to focus its efforts on returning humans to the Moon by 2020 and eventually sending them to Mars and "worlds beyond."[7] (Twelve U.S. astronauts walked on the Moon between 1969 and 1972. No humans have visited Mars.) The Vision also directed NASA to return the space shuttle to flight status following the February 2003 *Columbia* disaster; to complete construction of the International Space Station (ISS) in accord with existing international commitments; and to conclude U.S. participation in the ISS by the end of 2015. The first post-*Columbia* shuttle flight was launched in July 2005. The other goals remain to be accomplished.

To advise NASA on implementation of the Vision, President Bush established a Commission on the Implementation of U.S. Space Exploration Policy, chaired by Edward C. "Pete" Aldridge, Jr.[8] The Aldridge Commission issued its report in June 2004.[9] In April 2005,

[6] National Research Council, *America's Future in Space: Aligning the Civil Space Program with National Needs*, 2009, http://www.nap.edu/catalog/12701.html, pp. 46-48.

[7] President George W. Bush, speech at NASA headquarters, Washington, D.C., January 14, 2004,http:// history.nasa.gov/BushSEP.htm. For more information on the original Vision and initial reactions to it byCongress and the public, see CRS Report RS21720, *Space Exploration: Issues Concerning the Vision for SpaceExploration*, by Marcia S. Smith.

[8] Peter Aldridge was Under Secretary of Defense for Acquisition, Technology, and Logistics from 2001 to 2003. Hepreviously held other senior positions in both government and the aerospace industry. In the 1980s, he trained as ashuttle astronaut, but his mission was cancelled following the *Challenger* disaster in 1986.

NASA established an Exploration Systems Architecture Study (ESAS) to identify a strategy and technical architecture for implementing the Vision. The ESAS issued its final report in November 2005.[10] Since then, the reports of the Aldridge Commission and the ESAS have been the baseline for NASA's space exploration plans.

In the NASAAuthorization Act of 2005 (P.L. 109-155), Congress endorsed the Vision in broad terms and established several milestones for its implementation, including a statutory mandate to return to the Moon no later than 2020.[11] Nevertheless, it directed NASA to construct an architecture and implementation plan for its human exploration program "that is not critically dependent on the achievement of milestones by specific dates."[12]

The NASA Authorization Act of 2008 (P.L. 110-422) reaffirmed the Vision's broad goals, including the "eventual" return to the Moon and missions to other destinations in the solar system.[13] It expressed the sense of Congress that "America's friends and allies" should be invited to participate.[14] It directed NASA to take a "stepping stone approach" in which lunar exploration activities are designed and implemented with strong consideration to their future contribution to exploration beyond the Moon.[15] It directed that plans for a lunar outpost should not require its continuous occupation and that NASA should use commercial services for its lunar outpost activities "to the maximum extent practicable."[16]

Current Program to Implement the Vision

The current program for implementing the Vision addresses the conclusion of the space shuttle and International Space Station programs as well as the development and implementation of new vehicles for taking humans into Earth orbit and then back to the Moon. The major elements of the current program are as follows:

- Retire the space shuttle at the end of 2010 (or if necessary, in the first half of 2011). Rely on non-U.S. vehicles for human access to space until a replacement vehicle is developed.
- Terminate U.S. use of the International Space Station at the end of 2015.
- Under the Constellation program, develop new systems for space exploration:
  the Ares I rocket to launch astronauts into low Earth orbit, where the International Space Station is located;
  the Orion crew capsule, to be launched atop Ares I to carry astronauts into orbit and beyond;
  the Ares V heavy-lift rocket to send astronauts and equipment to the Moon; and
  the Altair lunar lander and various lunar surface systems.

---

[9] Commission on the Implementation of U.S. Space Exploration Policy, *A Journey to Inspire, Innovate, and Discover*, June 2004, http://www.nasa.gov/pdf/60736main_M2M_report_small.pdf.

[10] National Aeronautics and Space Administration, *NASA's Exploration Systems Architecture Study: Final Report*, NASA-TM-2005-214062, November 2005, http://www.nasa.gov/pdf/140649main_ESAS_full.pdf.

[11] P.L. 109-155, Section 101(b).

[12] P.L. 109-155, Section 503.

[13] P.L. 110-422, Section 402.

[14] P.L. 110-422, Section 401.

[15] P.L. 110-422, Section 403.

[16] P.L. 110-422, Section 404.

No FY2011 funds are currently budgeted for any shuttle flights that extend past the end of 2010. No FY2016 funds are currently budgeted for deorbiting the space station. The first crewed flight (or "initial operating capability") of Ares I and Orion is scheduled for early 2015. The first return to the Moon, using all the Constellation systems together, is planned for 2020, but NASA acknowledges that meeting that date will be difficult.

Issue for Congress: Cost and Schedule

Cost is likely to play a central role as congressional policy makers oversee the Vision's progress and considers proposals to modify it. During the Bush Administration, NASA stressed that its strategy was to "go as we can afford to pay," with the pace of the program set, in part, by the available funding.[17] The original plan in 2004 proposed adding a total of just $1 billion to NASA's budget for FY2005 through FY2009 to help pay for the Vision, with increases thereafter limited to the rate of inflation. Subsequent Administration budgets more than eliminated this increase, and actual appropriations by Congress were even less. As a result, most funding for the Vision has been redirected from other NASA activities, such as the planned termination of the space shuttle program.

NASA has not provided a cost estimate for the Vision as a whole. In 2004, it projected that developing capabilities for human exploration, not including robotic support missions, would cost a total of $64 billion up through the first human return to the Moon.[18] The Congressional Budget Office concluded that, based on historical trends, the actual cost could be much higher.[19] In its 2005 implementation plan, NASA estimated that returning astronauts to the Moon would cost $104 billion, not including the cost of robotic precursor missions or the cost of servicing the ISS after the end of the shuttle program.[20] In 2007, the Government Accountability Office (GAO) estimated the total cost for the Vision as $230 billion over two decades.[21] In April 2009, as directed in the 2008 authorization act, the Congressional Budget Office updated its 2004 budgetary analysis of the Vision. It found that NASA would need an additional $2 billion per year through FY2025 to keep the Vision activities on schedule, not counting probable cost growth in other activities.[22] In October 2009, the Augustine report stated that executing NASA's current plans would require an additional $3 billion per year, even with some schedule delays.[23]

Schedule is closely related to cost. For example, the 2009 CBO analysis found that NASA could maintain its currently planned budget by delaying its return to the Moon by approximately three years.[24] The tradeoffs can be difficult to quantify, however. The

[17] See, for example, Michael D. Griffin, Administrator, National Aeronautics and Space Administration, testimonybefore the Senate Committee on Commerce, Science, and Transportation, Subcommittee on Space, Aeronautics, andRelated Sciences, February 28, 2007, http://commerce.senate.gov/public/_files/Testimony_MichaelDGriffin_NASA_FY2008PostureStatementFINAL22707.pdf.

[18] Congressional Budget Office, *A Budgetary Analysis of NASA's New Vision for Space Exploration*, September 2004,http://www.cbo.gov/ftpdocs/57xx/doc5772/09-02-NASA.pdf, p. xi.

[19] Congressional Budget Office, *A Budgetary Analysis of NASA's New Vision for Space Exploration*, pp. xi-xiv.

[20] National Aeronautics and Space Administration, *Exploration Systems Architecture Study: Final Report*, p. 676.

[21] Government Accountability Office, *High Risk Series*, GAO-07-310, January 2007, p. 75.

[22] Congressional Budget Office, *The Budgetary Implications of NASA's Current Plans for Space Exploration*, April2009, http://www.cbo.gov/ftpdocs/100xx/doc10051/04-15-NASA.pdf, pp. 2-3 and 12-13. The statutory mandate forthis study was in P.L. 110-422, Section 410.

[23] Review of U.S. Human Spaceflight Plans Committee, *Seeking a Human Spaceflight Program Worthy of a GreatNation*, pp. 15-17.

[24] Congressional Budget Office, *The Budgetary Implications of NASA's Current Plans for Space Exploration*, pp. 2-3 and 7-9.

Augustine report, unlike the CBO analysis, found that under NASA's current budget plans, "human exploration beyond low-Earth orbit is not viable" and currently planned budgets would delay a return to the Moon "well into the 2030s, if ever."[25] Schedule delays are already evident. For example, the initial operating capability for Orion and Ares I was originally planned for 2012; it is now planned for 2015; the Augustine committee concluded that 2017 is more likely.

Issue for Congress: Why the Moon?

Ever since the Vision was first announced, some analysts have questioned its choice of the Moon as the headline destination for NASA's human exploration efforts. Some feel that revisiting the destination of the Apollo missions of 1969-1972 is a less inspiring goal than a new target would be.[26] Some doubt the scientific rationale, suggesting that robotic missions to the Moon could accomplish as much or more at lower cost and without risking human lives, or that more could be learned by visiting another destination that has been studied less by previous missions. Some are simply concerned about the cost.

Supporters counter that the Moon is the closest destination beyond Earth orbit and could serve as a stepping stone for subsequent destinations. As Earth's nearest neighbor, the Moon is of great scientific interest. Missions to the Moon would provide an opportunity to develop and test technologies and gain experience working in space. According to some advocates, the Moon might literally be a staging point for future missions. For some in Congress, concerned about national security or national prestige, the prospect of a manned Chinese mission to the Moon is a strong motivation to reestablish a U.S. presence. For many who have supported the Vision up to this point, completing it may have become important in itself; part of the Vision's original purpose was to set a goal for NASA that would give the agency direction and enhance its public support, and supporters may fear that changing plans at this point would weaken NASA, whether or not a better plan could be devised.

Issue for Congress: "The Gap" and Utilization of the Space Station

In order to fund the cost of the Vision and because of safety concerns following the *Columbia* disaster in 2003, NASA intends to end the space shuttle program once construction of the ISS is complete in 2010. The shuttle's successors, Orion and Ares I, are not expected to be ready for crewed flight until at least 2015. The difference between these dates is generally referred to as "the gap." Congressional policy makers and others have expressed concerns about U.S. access to space during the gap. The NASAAuthorization Act of 2005 declared it to be U.S. policy "to possess the capability for human access to space on a continuous basis."[27] Former NASA Administrator Michael Griffin, a strong advocate of the Vision, has referred to the gap as "unseemly in the extreme."[28]

---

[25] Review of U.S. Human Spaceflight Plans Committee, *Seeking a Human Spaceflight Program Worthy of a Great Nation*, pp. 17 and 84.

[26] For example, Apollo astronaut Buzz Aldrin has called returning to the Moon "not visionary." See Buzz Aldrin, "Time to Boldly Go Once More," *New York Times*, July 16, 2009.

[27] P.L. 109-155, Section 501.

[28] For example, see Michael D. Griffin, Administrator, National Aeronautics and Space Administration, testimony before the before the House Committee on Science and Technology, February 13, 2008, http://www.nasa.gov/pdf/ 211844main_House_Science_Committee_Oral_13_Fe08.pdf.

Under current plans, Russian spacecraft will be the only means of access to the ISS for humans during the gap. A variety of alternatives are being considered for cargo. These points are discussed further below in the section "Post-Shuttle Access to the ISS."

The prospect of the gap has intensified congressional concerns about whether the capabilities of the ISS will be fully utilized. In addition to the uncertainty about U.S. access to the ISS during the gap period, it appears possible that U.S. use of the station will end at about the same time as the shuttle's successors first become available, or even before.

Human Spaceflight: The Augustine Committee

The Review of U.S. Human Spaceflight Plans Committee was formally chartered on June 1, 2009. It was chaired by Norman R. Augustine, a former chairman and chief executive officer of Lockheed Martin Corporation and a member of the President's Council of Advisors on Science and Technology under Presidents of both parties. Other committee members included scientists, engineers, astronauts, educators, executives of established and new aerospace firms, former presidential appointees, and a retired Air Force general.[29] The committee reported jointly to the Administrator of NASA and the Director of the Office of Science and Technology Policy in the Executive Office of the President. The committee's charter defined its scope and objectives as follows:[30]

> The Committee shall conduct an independent review of ongoing U.S. human space flight plans and programs, as well as alternatives, to ensure the Nation is pursuing the best trajectory for the future of human space flight – one that is safe, innovative, affordable, and sustainable. The Committee should aim to identify and characterize a range of options that spans the reasonable possibilities for continuation of U.S. human space flight activities beyond retirement of the Space Shuttle. The identification and characterization of these options should address the following objectives: a) expediting a new U.S. capability to support utilization of the International Space Station (ISS); b) supporting missions to the Moon and other destinations beyond low-Earth orbit (LEO); c) stimulating commercial space flight capability; and d) fitting within the current budget profile for NASA exploration activities.[31]
>
> In addition to the objectives described above, the review should examine the appropriate amount of research and development and complementary robotic activities needed to make human space flight activities most productive and affordable over the long term, as well as appropriate opportunities for international collaboration. It should also evaluate what capabilities would be enabled by each of the potential architectures considered. It should evaluate options for extending ISS operations beyond 2016.

Options Identified by the Augustine Committee

The committee released its final report in October 2009. It identified five options: two within the current budget profile and three that would require about an additional $3 billion

---

[29] See *Meet the Committee*, http://www.nasa.gov/offices/hsf/members/index.html.

[30] *Charter of the Review of U.S. Human Space Flight Plans Committee,* http://www.nasa.gov/offices/hsf/about/charter.html.

[31] It was subsequently agreed that the committee would also consider options not constrained by the current budget profile, if necessary to satisfy the other objectives. Review of U.S. Human Spaceflight Plans Committee, *Seeking a Human Spaceflight Program Worthy of a Great Nation*, p. 7.

per year. In the committee's judgment, developing Ares I, Orion, and the other Constellation systems is likely to take longer than NASA currently plans, and the options presented by the committee reflect these expected delays. The options are as follows:[32]

- **Option 1: Current Budget, Current Program.** This option is the currentprogram, modified only to provide funds for space shuttle flights in FY2011 andfor deorbiting the International Space Station in FY2016. The first crewed flight of Ares I and Orion is no earlier than 2017, after the International Space Station has been deorbited. Ares V is not available until the late 2020s, and there are insufficient funds to develop Altair and the lunar surface systems needed for returning to the Moon until well into the 2030s, if ever.
- **Option 2: Current Budget, Extend Space Station, Explore Moon UsingAres V Lite.** This option extends use of the International Space Station to 2020 and begins a program of lunar exploration using a variant of Ares V known asAres V Lite. It develops commercial services to transport humans into low Earthorbit. It delivers a heavy-lift capability in the late 2020s, but it does not develop the other systems needed for returning to the Moon for at least the next twodecades.
- **Option 3: Additional Budget, Current Program.** Like Option 1, this option isthe current program, modified to provide funds for space shuttle flights inFY2011 and to deorbit the International Space Station in FY2016. The first crewed flight of Ares I and Orion would still be after the International SpaceStation is deorbited. The additional funding, however, would permit a humanlunar return in the mid-2020s.
- **Option 4: Additional Budget, Extend Space Station, Explore Moon First.** Like Option 2, this option extends use of the International Space Station to 2020 and uses commercial services to transport humans into low Earth orbit. The first destination beyond Earth orbit is still the Moon. There are two variants to this option. **Variant 4A** develops the Ares V Lite for lunar exploration as in Option 2. **Variant 4B** extends the space shuttle program to 2015 and develops a heavy-lift vehicle for lunar missions that is more directly shuttle-derived. Both variants permit a human lunar return by the mid-2020s.
- **Option 5: Additional Budget, Extend Space Station, Flexible Path for Exploration.** Like Option 4, this option extends use of the International Space Station to 2020 and uses commercial services to transport humans into low Earth orbit. Missions beyond Earth orbit, however, follow a "flexible path" of increasingly distant destinations—such as lunar fly-bys, rendezvous with asteroids and comets, and Mars fly-bys—without initially attempting a lunar landing. A lunar landing would be possible by the mid to late 2020s. **Variant 5A** employs the Ares V Lite. **Variant 5B** uses a commercial heavy-lift rocket derived from the Evolved Expendable Launch Vehicle (EELV). **Variant 5C** develops a shuttle-derived vehicle for heavy lift as in Variant 4B. (These alternative launch vehicles are discussed further later in this chapter.)

[32] Review of U.S. Human Spaceflight Plans Committee, *Seeking a Human Spaceflight Program Worthy of a Great Nation*, pp. 15-16.

Although the committee did not recommend any particular one of these options over the others in its report, it made a number of findings and comments that put the options into context:[33]

- Option 1 and Option 2 fit within the current budget profile, but "neither allowsfor a viable exploration program. In fact, the Committee finds that no plancompatible with the FY2010 budget profile permits human exploration tocontinue in any meaningful way." The additional funding contemplated in Options 3, 4, and 5 is necessary for "an exploration program that will be a source of pride for the nation."
- "The return on investment to both the United States and our international partners would be significantly enhanced by an extension of the life of the [International Space Station]. A decision not to extend its operation would significantly impair U.S. ability to develop and lead future international spaceflight partnerships."
- Commercial services to launch crews into Earth orbit "are within reach. Whilethis presents some risk, it could provide an earlier capability at lower initial and life-cycle costs than the government could achieve."
- Of the heavy-lift alternatives, Ares V Lite is "the most capable." The commercial EELV derivative "has an advantage of potentially lower operating costs, but requires significant restructuring of NASA" including "a different (and significantly reduced) role." A shuttle-derived vehicle would "take maximum advantage of existing infrastructure, facilities, and production capabilities."
- Variant 4B, which extends operation of the space shuttle to 2015, is "the onlyforeseeable way to eliminate the gap in U.S. human-launch capability."
- "Mars is the ultimate destination for human exploration of the inner solar system; but it is not the best first destination. Visiting the 'Moon First' and following the 'Flexible Path' are both viable exploration strategies. The two are not necessarily mutually exclusive; before traveling to Mars, we could extend our presence in free space and gain experience working on the lunar surface."

Questions for Congressional Policy Makers to Consider

The Augustine committee identified five questions "that could form the basis of a plan for U.S. human spaceflight":[34]

- What should be the future of the space shuttle?
- What should be the future of the International Space Station?
- On what should the next heavy-lift launch vehicle be based?
- How should crew be carried to low Earth orbit?
- What is the most practicable strategy for exploration beyond low Earth orbit?

These five questions focus on designing a future program of human spaceflight. In keeping with the committee's charter, the questions do not address NASA's other programs,

[33] Review of U.S. Human Spaceflight Plans Committee, *Seeking a Human Spaceflight Program Worthy of a Great Nation*, pp. 16-17.

[34] Review of U.S. Human Spaceflight Plans Committee, *Seeking a Human Spaceflight Program Worthy of a Great Nation*, p. 22.

and they take it as given that a human spaceflight program *should* be implemented. Congress may therefore wish to consider additional questions such as these:

- Is human spaceflight beyond low Earth orbit worth the cost and risk?
- If not, are there alternatives that would accomplish some of the same goals?
- What is the future of NASA's other activities, such as robotic exploration,science, and aeronautics research?

Each of these issues is discussed in more detail later in this chapter.

Human Spaceflight: Administration Proposals

In its FY2011 budget request, the Obama Administration proposed cancelling the Constellation program and eliminating the return of humans to the Moon as NASA's primary goal.[35] Instead, NASA would encourage the private sector to develop commercial space transportation services to carry astronauts to and from the International Space Station. For spaceflight beyond Earth orbit, NASA would emphasize long-term technology development rather than near-term development of specific flight systems. Operation of the International Space Station would continue until at least 2020. When asked about destinations for future human exploration of space, NASA officials stated that Mars would be the ultimate goal, but that other intermediate destinations would come first. They described these proposals as consistent with the "Flexible Path" option identified by the Augustine committee.

Congressional and Public Reaction

Congress and the public at large reacted mostly negatively to the Administration's proposals. Their concerns included the potential negative impact of Constellation's cancellation on employment in the aerospace industry, the lack of a specific destination and schedule to replace the goal of returning humans to the Moon, and the risk that the private sector might not in fact develop commercial space transportation services that meet NASA's needs. In the media, attention focused on the proposed cancellation of Constellation, with less notice of the programs that would replace it, such as increased technology development and stimulation of commercial space transportation services. Press accounts often reported the Administration's proposals as cutting NASA's budget, or eliminating its human spaceflight program, even though the proposed FY2011 budget for NASA was actually an increase over previous plans and included other human spaceflight activities to replace Constellation.

Supporters of Constellation have been particularly concerned about its status during FY2010. The FY2010 appropriations act prohibits NASA from using FY2010 or prior-year funds to terminate or eliminate "any program, project, or activity of the architecture for the Constellation program" or to create or initiate any new program, project, or activity.[36] Some analysts and policy makers have expressed concern that NASA contracting decisions and

[35] For more information on the FY2011 budget request for NASA, see CRS Report R41161, *Commerce, Justice, Science, and Related Agencies: FY2011 Appropriations*.

[36] P.L. 111-117, Division B, Title III.

other actions during FY2010 may be in violation of the appropriations provision.[37] NASA officials reply that they are continuing to implement the Constellation program during FY2010 in full compliance with the law, even though they intend to terminate the program in FY2011.

Modifications to the Administration Proposals

On April 15, 2010, President Obama gave a speech at the Kennedy Space Center in Florida that attempted to answer some of these public and congressional concerns.[38] In this speech, he announced several modifications to the original FY2011 budget request proposals:

- Development of a modified Orion crew capsule would continue. The modifieddesign would provide an emergency escape capability for the International SpaceStation, however, rather than transporting crews to and from the station on aregular basis.
- The next human mission beyond Earth orbit would be to an asteroid and wouldtake place in 2025. This would be the first human mission to a destination moredistant than the Moon. Subsequent missions to orbit Mars would take place in themid-2030s. A human landing on Mars would remain the ultimate goal.
- NASA's increased technology efforts would focus on the development of a newheavy-lift rocket, with a decision in 2015 on a specific heavy-lift architecture forexploration of deep space.
- Independent of the FY2011 budget, an Administration task force will address economic development and the aerospace industry in the region of Florida knownas the Space Coast.

It remains to be seen whether Congress and the public will receive these changes to the Administration's original proposals favorably.

Science

About two-thirds of NASA's budget is associated with human spaceflight. Most of the rest is devoted to unmanned science missions. These science missions fall into four categories: Earth science, planetary science, heliophysics, and astrophysics. The latter three are sometimes known collectively as space science.

In part because of concerns about climate change, both Congress and the Administration have recently placed increased emphasis on Earth science. In the FY2006 and FY2007 budget cycles NASA had no separate budget for Earth science, and supporters became concerned that this was adversely affecting the field. In late 2006, NASA reorganized the Science Mission

[37] See, for example, the letter from 27 Members of Congress to NASA Administrator Charles Bolden, February 12, 2010, online at http://www.posey.house.gov/UploadedFiles/LetterToBolden-CancellingConstellation-Feb15-2010.pdf.

[38] For a transcript of the speech, see "Remarks by the President on Space Exploration in the 21st Century," White House press release, April 15, 2010, http://www.whitehouse.gov/the-press-office/remarks-president-space-exploration-21stcentury. See also two White House fact sheets provided in association with the speech: "A Bold Approach for Space Exploration and Discovery," http://www.whitehouse.gov/sites/default/files/microsites/ostp/ostp-space-conffactsheet.pdf, and "Florida's Space Workers and the New Approach to Human Spaceflight," http://www.whitehouse.gov/sites/default/files/microsites/ostp/nasa-space-conf-factsheet.pdf.

Directorate, creating a separate Earth Science Division. The National Research Council recommended in early 2007 that the United States "should renew its investment in Earth observing systems and restore its leadership in Earth science and applications."[39] In response, Congress and the Administration increased the share of NASA's science funding devoted to Earth science from 26% in FY2008 to 32% in FY2010. In addition, NASA allocated 81% of the science funding it received under the American Recovery and Reinvestment Act of 2009 (P.L. 111-5) to Earth science. The Administration's FY2011 budget would provide substantial increases for Earth science funding, including a five-year, $2.1 billion global climate initiative.

In recent years, Congress has sought to ensure that NASA's science program includes a balanced variety of approaches to R&D rather than focusing only on certain types of missions. For example, the NASA Authorization Act of 2008 stated that the science program should include space science missions of all sizes as well as mission-enabling activities such as technology development, suborbital research, and research and analysis (R&A) grants to individual investigators.[40] According to the National Research Council, "practically all relevant external advisory reports have emphasized the importance of mission-enabling activities," but determining their proper scale has been challenging "throughout NASA's history."[41] In the past few years, funding for planetary science technology has increased significantly, but funding for Earth science technology has increased only slightly; the astrophysics and heliophysics programs do not have dedicated technology subprograms. Funding for suborbital rocket operations increased from $51 million in FY2008 to $66 million in FY2010, but the trend is unclear as the latter amount was down from $77 million in FY2009. Funding for R&A grants, which NASA controversially proposed to reduce significantly as recently as FY2007, has recovered as the result partly of the Administration's own initiatives and partly of congressional action on appropriations legislation.

In December 2009, the National Research Council recommended ways to make the mission-enabling activities of NASA's science programs more effective through more active management. These recommendations included establishing explicit objectives and metrics, making budgets more transparent, and clearly articulating the relationships between mission-enabling activities and the ensemble of missions they are intended to support.[42] The NASA Authorization Act of 2008 stated that the technology development program should include long-term activities that are "independent of the flight projects under development."[43] NASA may sometimes find it challenging to balance this independence against the goal of linking mission-enabling activities to the missions they support.

NASA's science programs have a history of periodic review by the National Academies. Such reviews typically take place every 10 years, so they are commonly known as decadal surveys. The NASA Authorization Act of 2005 mandated an Academy review of each division of NASA's science directorate every five years.[44] The NASAAuthorization Act of 2008 also mandated periodic reviews and directed that they include independent estimates of

---

[39] National Research Council, *Earth Science and Applications from Space: National Imperatives for the Next Decade and Beyond*, 2007, http://www.nap.edu/catalog/11820.html.

[40] P.L. 110-422, Section 504.

[41] National Research Council, *An Enabling Foundation for NASA's Earth and Space Science Missions*, December2009, http://www.nap.edu/catalog/12822.html, p. vii.

[42] National Research Council, *An Enabling Foundation for NASA's Earth and Space Science Missions*, p. 4.

[43] P.L. 110-422, Section 501.

[44] P.L. 109-155, Section 301.

the cost and technical readiness of each mission assessed.[45] Decadal surveys by the National Academies are generally well received by NASA and are widely respected in the science and science policy communities. On the other hand, the expertise of the National Academies is primarily scientific. It is unclear whether their analysis of mission cost and readiness will be considered equally authoritative.

Aeronautics

After human spaceflight and science, NASA's largest activity is research on aeronautics, the science and technology of flight within Earth's atmosphere. There is a history of disagreement in Congress about the appropriate role of this program. Supporters argue that the aviation industry is vital to the economy, especially because aircraft are a major component of U.S. exports. They claim that government funding for aeronautics research can contribute to U.S. competitiveness and is necessary in light of similar programs in Europe and elsewhere.[46] Opponents counter that the aviation industry itself should pay for the R&D it needs. Against the background of this debate, NASA aeronautics programs have focused increasingly on long-term fundamental R&D and on research topics with clear public purposes, such as reducing noise and emissions, improving safety, and improving air traffic control.

In 2005, Congress directed the President to develop a national policy for aeronautics R&D.[47] The National Science and Technology Council (NSTC), part of the Executive Office of the President, issued this policy in December 2006.[48] The policy established general principles and goals for federal aeronautics activities, laid out the roles and responsibilities of NASA and other agencies, and directed the NSTC to issue a national aeronautics R&D plan at least every two years. The NSTC released the first national aeronautics R&D plan in December 2007.[49] It released a draft update for public comment in November 2009.[50] The NASA Authorization Act of 2008 stated that NASA's aeronautics research program should be "guided by and consistent with" the national aeronautics R&D policy.[51]

---

[45] P.L. 110-422, Section 1104.

[46] In 2005, a NASA-funded report found that European government support for aeronautics R&D was growing and that European countries "support civil aeronautics research on the basis of industrial policies." (Hans J. Weber et al., "Study of European Government Support to Civil Aeronautics R&D," August 15, 2005, http://www.aeronautics.nasa.gov/docs/ tecop_europe_aero_r&d.pdf.) The report also found that European aeronautics R&D was increasingly guided by a 2001 European Union report that called for European countries to invest about $100 billion in the topic over 20 years. See *European Aeronautics: A Vision for 2020*, http://www.acare4europe.org/docs/Vision%202020.pdf.

[47] P.L. 109-155, Section 101(c). A similar but less detailed provision was previously included in Section 628 of the Science, State, Justice, Commerce, and Related Agencies Appropriations Act, 2006 (P.L. 109-108).

[48] Executive Office of the President, National Science and Technology Council, *National Aeronautics Research and Development Policy*, December 2006, http://www.aeronautics.nasa.gov/releases/ national_aero nautics_rd_policy_dec_2006.pdf.

[49] Executive Office of the President, National Science and Technology Council, *National Plan for AeronauticsResearch and Development and Related Infrastructure*, December 2007, http://www.a eronautics.nasa.gov/releases/aero_rd_plan_final_21_dec_2007.pdf.

[50] Executive Office of the President, National Science and Technology Council, *National Aeronautics Research andDevelopment Plan Biennial Update*, Draft 3.0, November 2009, http://www.ostp.gov/aeroplans/pdf/ nat_aero_rd_plan_public_comments.pdf.

[51] P.L. 110-422, Section 301.

In June 2006, in response to a congressional mandate, the National Research Council of the National Academies released a decadal strategy for federal civil aeronautics activities, with a particular emphasis on NASA's aeronautics research program.[52] Along with other recommendations, the report identified 51 technology challenges to serve as the foundation for aeronautics research at NASA for the next decade. In the 2008 authorization act, Congress directed NASA to align its fundamental aeronautics research program with these technology challenges "to the maximum extent practicable within available funding" and to increase the involvement of universities and other external organizations in that program.[53] It also mandated periodic Academy reviews of the NASA aeronautics program and directed that they include independent estimates of the cost and technical readiness of each mission assessed.[54] As noted above with respect to its decadal surveys of NASA science, while the National Academies are widely respected for their scientific expertise, it is unclear whether their analysis of cost and technical readiness will be considered equally authoritative.

The aeronautics program's heavy use of shared facilities and capabilities, such as wind tunnels and supercomputers, has sometimes created challenges. For example, when NASA introduced full-cost accounting in the FY2004 budget request, the stated cost of the aeronautics program increased significantly because facility costs had previously been budgeted in another account. At least partly in response to these concerns, NASA subsequently established a separate Aeronautics Test Program in the aeronautics directorate and a Strategic Capabilities Assets Program outside the directorate. It has also sometimes been difficult for NASA to balance its stewardship of unique aeronautics facilities, often used by other agencies and by industry as well as by NASA itself, against the cost of maintaining those facilities. In 2005, Congress directed NASA to establish a separate account to fund aeronautics test facilities, to charge users of NASA test facilities at a rate competitive with alternative facilities, and not to implement a policy seeking full cost recovery for a facility without giving 30 days' notice to Congress.[55] To accompany the national aeronautics R&D plan, the Aeronautics Science and Technology Subcommittee of the NSTC is developing a national aeronautics research, development, test, and evaluation infrastructure plan. This infrastructure plan is scheduled for completion in 2010.[56]

There is ongoing congressional interest in the relationship between NASA's aeronautics program and related efforts by the Federal Aviation Administration (FAA) and the Department of Defense (DOD). One aspect of this relationship is the interagency Joint Planning and Development Office (JPDO), which oversees the development of a Next Generation Air Transportation System (NGATS) for improved airspace management.[57] Congress has directed NASA to align the Airspace Systems program of its Aeronautics Research Directorate with the objectives of the JPDO and NGATS.[58]

---

[52] National Research Council, *Decadal Survey of Civil Aeronautics: Foundation for the Future*, 2006, http://www.nap.edu/catalog/11664.html. The congressional mandate was in P.L. 109-155, Section 421(c).

[53] P.L. 110-422, Section 303.

[54] P.L. 110-422, Section 1104.

[55] P.L. 109-155, Section 205.

[56] Executive Office of the President, National Science and Technology Council, webpage of the Aeronautics Science and Technology Subcommittee, http://www.ostp.gov/aeroplans/index.htm.

[57] The establishment of JPDO was mandated by Section 709 of the Vision 100 – Century of Aviation Reauthorization Act (P.L. 108-176).

[58] P.L. 109-155, Section 423.

Education

In 2008, a congressionally mandated National Academies review of NASA education programs found that even though NASA is uniquely positioned to interest students in science, technology, and engineering, its education programs are not as effective as they could be.[59] The report found that NASA has no coherent plan to evaluate its education programs, and few of them have ever been formally evaluated. It recommended that NASA develop an evaluation plan and use the results of the evaluations to inform project design and improvement. It found that the operating directorates, rather than the Office of Education, fund about half of the agency's primary and secondary education activities. It recommended that the Office of Education focus on coordination and oversight, including advocacy for the inclusion of education activities in the programs of the operating directorates. Congress directed NASA to prepare a plan by October 2009 in response to the recommendations of the National Academies, including a schedule and budget for any actions that have not yet been implemented.[60] As of December 2009, the plan had not yet been completed.[61]

Unlike the Department of Education or the National Science Foundation, NASA does not have a lead role in federal education programs. As a result, some analysts may view NASA's education activities as secondary to its primary efforts in spaceflight, science, and aeronautics. Congress, however, is typically supportive of NASA education programs and often provides more funding for them than NASA requests. This imbalance between Administration and congressional priorities, the dispersed nature of NASA's education activities outside the Office of Education, and the tendency for congressional funding increases to be dedicated to specific one-time projects rather than to ongoing programs, may make it difficult for NASA to plan and manage a coherent, unified education program.

Balancing Competing Priorities

Ever since the announcement of the Vision, NASA's emphasis on exploration has created concerns about the balance between human spaceflight and NASA's other activities, especially science and aeronautics. Because most funding for the Vision has been redirected from other NASA activities, advocates of science and aeronautics have feared that their programs will be cut in order to pay for human exploration activities. Congress, while fully supporting the Vision, has been clear about the need for balance.The NASAAuthorizationAct of 2005 directed NASA to carry out "a balanced set of programs," including human space flight in accordance with the Vision, but also aeronautics R&D and scientific research, the latter to include robotic missions and research not directly related to human exploration.[62] TheNASA Authorization Act of 2008 found that NASA "is and should remain a multimission agency with a balanced and robust set of core missions in science, aeronautics, and human space flight and exploration" and "encouraged" NASA to coordinate its exploration activities

[59] National Research Council, *NASA's Elementary and Secondary Education Program: Review and Critique*, 2008, http://www.nap.edu/catalog/12081.html. The mandate for this review was in P.L. 109-155, Section 614.

[60] P.L. 110-422, Section 701.

[61] NASA Office of Legislative Affairs, personal communication, December 16, 2009.

[62] P.L. 109-155, Section 101(a)(1).

with its science activities.[63] In January 2010, NASA Administrator Charles Bolden assured a group of scientists that "the future of human spaceflight will not be paid for out of the hide of our science budget."[64]

Balancing these competing priorities depends on answering questions, raised earlier in this chapter, about NASA's purpose. More than 50 years ago, President Eisenhower's advisors were aware that a space program was justified both by "the compelling urge of man to explore and to discover" and by "scientific observation and experiment which will to add to our knowledge and understanding." Today, there is still no consensus about how to balance these purposes. Some policy makers believe that a space program can best be justified by tangible benefits to economic growth and competitiveness. Others believe that its most important role is to be a source of national pride, prestige, and inspiration.

Space Shuttle Program

Since its first launch in April 1981, the space shuttle has been the only U.S. vehicle capable of carrying humans into space. After a few remaining flights during 2010 (some may slip into early 2011) current plans call for the space shuttle program to end. Although some advocates and policy makers would like to extend the program, technical and management issues are making that ever more difficult as the scheduled termination approaches. Congress's attention is increasingly on managing the transition of the shuttle workforce and facilities and on addressing the projected multi-year gap in U.S. access to space between the last shuttle flight and the first flight of its successor.

Why the Shuttle Program Is Ending

The oldest shuttle is approaching 30 years old; the youngest is approaching 20. Although many shuttle components have been refurbished and upgraded, the shuttles as a whole are aging systems. Most analysts consider the shuttle design to be based, in many respects, on obsolete or obsolescent technology. The original concept of the shuttle program was that a reusable launch vehicle would be more cost-effective than an expendable one, but many of the projected cost savings depended on a flight rate that has never been achieved. Over the years, NASA has attempted repeatedly, but unsuccessfully, to develop a second-generation reusable launch vehicle to replace the shuttle. In 2002, NASA indicated that the shuttle would continue flying until at least 2015 and perhaps until 2020 or beyond.

The *Columbia* disaster in 2003 forced NASA to revise that plan. Within hours of the loss of the space shuttle *Columbia* and its seven astronauts, NASA established the Columbia Accident Investigation Board to determine the causes of the accident and make recommendations for how to proceed.[65] The board concluded that the shuttle "is not

[63] P.L. 110-422, Sections 2 and 409.

[64] NASA Administrator Charles F. Bolden, Jr., address to a meeting of the American Astronomical Society, January 5, 2010, http://www.nasa.gov/pdf/415511main_Bolden_AAS_Remarks_010510.pdf.

[65] For more details, see CRS Report RS21606, *NASA's Space Shuttle Columbia: Synopsis of the Report of the ColumbiaAccident Investigation Board*, by Marcia S. Smith.

inherently unsafe" but that several actions were necessary "to make the vehicle safe enough to operate in the coming years."[66] It recommended 15 specific actions to be taken before returning the shuttle to flight. In addition, it found that

> because of the risks inherent in the original design of the space shuttle, because the design was based in many aspects on now-obsolete technologies, and because the shuttle is now an aging system but still developmental in character, it is in the nation's interest to replace the shuttle as soon as possible as the primary means for transporting humans to and from Earth orbit.[67]

The board recommended that if the shuttle is to be flown past 2010, NASA should "develop and conduct a vehicle recertification at the material, component, subsystem, and system levels" as part of a broader and "essential" Service Life Extension Program.[68]

The announcement of the Vision for Space Exploration in 2004 created another reason to end the shuttle program: money. Before the shuttle program began to ramp down, it accounted for about 25% of NASA's budget. Making those funds available for the Vision became a primary motivation for ending the program.

Possible Extension of the Shuttle Program

Despite the safety risks identified by the Columbia Accident Investigation Board and the need to reallocate the shuttle's funding stream to other purposes, some policy makers and advocates remain eager to extend the program. For example, the American Space Access Act (H.R. 1962) would extend the program to 2015, and the NASAAuthorization Act of 2008, passed shortly before the 2008 presidential election, directed NASA not to take any action that would preclude the new President from deciding to extend the shuttle program past 2010.[69] One of the options put forward by the Augustine committee (Variant 4B) would include extending the shuttle program to 2015.

A decision to extend the program would create challenges relating to cost, schedule, and safety. With the planned termination date approaching, some contracts for shuttle components have already run out, and some contractor personnel have already been let go.[70] Reestablishing the capability to operate the program would likely incur costs and delays, and this potential will grow as the planned termination date approaches. The recertification process recommended by the Columbia Accident Investigation Board could be costly and time-consuming, although the board itself gave no estimate of either cost or schedule. At this point, completing a recertification in time to maintain a continuous flight schedule might already be difficult. Congressional policy makers or the Administration could simply decide to continue flying anyway, in parallel with the recertification process—in effect, NASA has already done this to some extent with the decision to allow a few flights to slip into 2011 if

---

[66] Columbia Accident Investigation Board, untitled report, August-October 2003, http://caib.nasa.gov/, vol. 1, p. 208.

[67] Columbia Accident Investigation Board, vol. 1, pp. 210-211.

[68] Columbia Accident Investigation Board, vol. 1, p. 209.

[69] P.L. 110-422, Section 611(d).

[70] For example, see Tariq Malik, "NASA Begins Job Cuts for Shuttle Retirement," Space.com, May 1, 2009.

necessary—but policy makers could suffer political repercussions from such a choice if another serious accident occurred.

During the 2009 presidential transition, the GAO identified the pending retirement of the space shuttle in 2010 as one of 13 "urgent issues" facing the incoming Obama Administration.[71] The GAO also stated that "according to NASA, reversing current plans and keeping the shuttle flying past 2010 would cost $2.5 billion to $4 billion per year."[72]

Transition of Shuttle Workforce and Facilities

The transition of assets and personnel at the end of the shuttle program is of great interest to many in Congress and represents a major challenge for NASA. The shuttle workforce is a reservoir of unique expertise and experience that would be difficult for NASA and its contractors to reassemble once dispersed. NASA managers are particularly concerned to maintain key human spaceflight expertise and capabilities through the expected gap period before the first flight of the shuttle's successor. In certain communities, the loss of the shuttle workforce will have a significant economic impact. For individuals, the loss of specialized, well-paid employment that has been relatively stable for many years can be especially disruptive at a time when the job market is already unusually difficult. Finding the best alternative use of facilities and equipment is important for getting the best value for the taxpayer.

NASA's transition management plan, issued in August 2008, establishes a timeline for the post-shuttle transition, defines organizational responsibilities for various aspects of the transition, establishes goals and objectives, and outlines planning and management challenges such as management of human capital and disposition of infrastructure.[73] As it notes, the scope of the transition is huge:

> The SSP [space shuttle program] has an extensive array of assets; the program occupies over 654 facilities, uses over 1.2 million line items of hardware and equipment, and employs over 2,000 civil servants, with more than 15,000 work year equivalent personnel employedbythe contractors. In addition, the SSP employs over 3,000 additional indirect workers through Center Management and Operations and service accounts. The total equipment acquisition value is over $12 billion, spread across hundreds of locations. The total facilities replacement cost is approximately $5.7 billion, which accounts for approximately one-fourth of the value of the Agency's total facility inventory. There are over 1,200 active suppliers and 3,000 to 4,000 qualified suppliers geographically located throughout the country.[74]

Congress has addressed a number of these issues through legislation:

---

[71] http://www.gao.gov/transition_2009/urgent/

[72] http://www.gao.gov/transition_2009/urgent/space-shuttle.php

[73] National Aeronautics and Space Administration, *NASA Transition Management Plan for Implementing the U.S. Space Exploration Policy*, JICB-001, August 2008, http://www.nasa.gov/pdf/202388main_Transition_Mgmt_Plan-Final.pdf.

[74] NASA, *Transition Management Plan*, pp. 7-8.

- In the NASA Authorization Act of 2005, Congress directed NASA to use thepersonnel, capabilities, assets, and infrastructure of the shuttle program "to thefullest extent possible consistent with a successful development program" indeveloping the vehicles now known as Orion, Ares I, and Ares V. It also requiredthe development of a transition plan for personnel affected by the termination ofthe shuttle program.[75]
- In the Commerce, Justice, Science, and Related Agencies Appropriations Act,2008, Congress directed NASA to prepare a strategy, to be updated at least everysix months, for minimizing job losses as a result of the transition from the shuttleto its successor.[76] The strategy report was first issued in March 2008 and was updated in October 2008 and July 2009.[77] As well as strategic information, it provides annual workforce projections for each NASA center and a summary of recent relevant actions by NASA and its contractors.
- In the NASA Authorization Act of 2008, Congress directed NASA to submit a plan for the disposition of the shuttles and associated hardware and to establish a Space Shuttle Transition Liaison Office to assist affected communities.[78] It provided for temporary continuation of health benefits for personnel whose jobs are eliminated as a result of the termination of the program.[79] It directed NASA to analyze the facilities and personnel that will be made available by the termination of the shuttle program and to report on other current and future federal programs that could use them.[80] The resulting report summarized the "mapping" process that NASA is using to align the civil servant and contractor shuttle workforce and the shuttle facilities at each NASA center with the needs of other programs.[81]
- In the Commerce, Justice, Science, and Related Agencies Appropriations Act, 2010, and in previous NASA appropriations acts for several years, Congress prohibited NASA from using appropriated funds to implement reductions in force (RIFs) or other involuntary separations, except for cause.[82]

Before the release of the Administration's FY2011 budget, many of the personnel currently employed in the shuttle program were expected to transition to the Constellation program. The proposed cancellation of Constellation introduces new uncertainty into these plans.

[75] P.L. 109-155, Section 502.

[76] Consolidated Appropriations Act of 2008 (P.L. 110-161), Division B.

[77] For the most recent update, see *NASA Space Shuttle Workforce Transition Strategy Pursuant to FY 2008 Consolidated Appropriations Act (P.L. 110-161), July 2009 Update*, http://www.nasa.gov/pdf/372110main_7-2109%20Workforce%20Transition%20Strategy%203rd%20Edition.pdf.

[78] P.L. 110-422, Section 613.

[79] P.L. 110-422, Section 615.

[80] P.L. 110-422, Section 614.

[81] National Aeronautics and Space Administration, *Aerospace Skills Retention and Investment Reutilization Report*, July 2009.

[82] Consolidated Appropriations Act, 2010 (P.L. 111-117), Division B.

International Space Station

Construction of the International Space Station (ISS) began in 1998. The ISS is composed of crew living space, laboratories, remote manipulator systems, solar arrays to generate electricity, and other elements. Launched separately, these elements were assembled in space. Rotating crews have occupied the ISS, each for a period of four to six months, since November 2000. Construction continues, with an expected completion date in 2010 or perhaps early 2011.[83]

The framework for international cooperation on the ISS is the Intergovernmental Agreement on Space Station Cooperation, which was signed in 1998 by representatives of the United States, Russia, Japan, Canada, Belgium, Denmark, France, Germany, Italy, the Netherlands, Norway, Spain, Sweden, Switzerland, and the United Kingdom. The intergovernmental agreement has the status of an executive agreement in the United States, but is considered a treaty in all the other partner countries. It is implemented through memoranda of understanding between NASA and its counterpart agencies: the Russian Federal Space Agency (Roskosmos), the Japanese Aerospace Exploration Agency (JAXA), the Canadian Space Agency (CSA), and the European Space Agency (ESA).[84] The United States also has an ISS participation agreement with Brazil, independent of the 1998 framework.

Because of cost growth and schedule delays, the scope and capabilities of the ISS have repeatedly been downsized.[85] The original concept was not just a laboratory, but also an observatory; a transportation node; a facility for servicing, assembly, and manufacturing; and a storage depot and staging base for other missions.[86] By 1989, only the laboratory function remained, and even that was smaller and less capable than in the original plans. In 1993, Russia joined the space station partnership, a move that added foreign policy objectives to the program's goals. By 2001, following further downsizing, NASA saw three goals for the station: conducting world-class research, establishing a permanent human presence in space, and "accommodation of all international partner elements."[87] Following the announcement of the Vision in 2004, learning to live and work in space became a key justification for the ISS program, and ISS research was to be focused on the long-term effects of space travel on human biology.

Concerned that the station's function as a research laboratory was being eroded, Congress took several legislative actions. The NASAAuthorizationAct of 2005 required NASA to allocate at least 15% of the funds budgeted to ISS research to "life and microgravity science research that is not directly related to supporting the human exploration program."[88] It also

---

[83] For more details, see the ISS website, http://www.nasa.gov/mission_pages/station/main/index.html.

[84] The text of the bilateral memoranda of understanding can be found at http://www.nasa.gov/mission_pages/station/structure/elements/partners_agreement.html.

[85] For more information on the evolution of the space station's purposes and capabilities, see Marcia S. Smith,Congressional Research Service, "NASA's Space Station Program: Evolution of its Rationale and Expected Uses,"testimony before the Senate Committee on Commerce, Science, and Transportation, Subcommittee on Science andSpace, April 20, 2005, http://commerce.senate.gov/pdf/smith.pdf.

[86] James Beggs, Administrator, National Aeronautics and Space Administration, testimony before the HouseCommittee on Appropriations, Subcommittee on HUD-Independent Agencies, March 27, 1984.

[87] Daniel Goldin, Administrator, National Aeronautics and Space Administration, testimony before the HouseCommittee on Science, April 25, 2001.

[88] P.L. 109-155, Section 204.

required NASA to submit a research plan for utilization of the ISS.[89] Issued in June 2006, the plan described proposed R&D and utilization activities in each of six disciplinary areas.[90] It characterized the ISS as a long-duration test-bed for future lunar missions; a flight analog for future missions to Mars; a laboratory for research directly related to human space exploration, such as human health countermeasures, fire suppression, and life support; and an opportunity to gain experience in managing international partnerships for long-duration space missions. The plan stated that research not related to exploration would continue "at a reduced level." At about the same time, the National Academies issued a review of NASA's plans for the ISS.[91] This review noted "with concern" that the objectives of the ISS "no longer include the fundamental biological and physical research that had been a major focus of ISS planning since its inception." It concluded that "once lost, neither the necessary research infrastructure nor the necessary communities of scientific investigators can survive or be easily replaced."

The 2005 authorization act designated the U.S. portion of the ISS as a national laboratory, to be available for use by other federal agencies and the private sector.[92] As required by the act, NASA submitted a plan for this designation in May 2007.[93] It concluded that NASA use of the ISS must continue to have first priority, that use by non-NASA entities should be funded by those entities, and that "the availability of cost-effective transportation services will directly affect the ability of the ISS to operate as a national laboratory in the years to come." The impact that the national laboratory designation would have was initially unclear. In the NASAAuthorization Act of 2008, Congress directed NASA to establish an advisory committee on the effective utilization of the ISS as a national laboratory.[94] As of mid-2009, NASA had established agreements for use of the ISS with at least five other federal agencies, three private firms, and one university, and had identified "firm interest" in using the ISS for education; human, plant, and animal biotechnologies; aerospace technologies; and defense sciences research.[95] NASA officials believe that about half of planned U.S. utilization resources on the ISS could be available for non-NASA use.[96]

When the space station was first announced, its assembly was to be complete by 1994. In 1998, when construction actually began, it was expected to be complete by 2002, with operations through at least 2012. Completion is now scheduled for 2010 or perhaps early 2011. As recently as 2003, NASA briefing charts showed operations possibly continuing through 2022. Under the Vision, U.S. utilization is scheduled to end after 2015, but widespread efforts to extend that date are ongoing.

---

[89] P.L. 109-155, Section 506.

[90] National Aeronautics and Space Administration, *Research and Utilization Plan for the International Space Station*, June 2006, http://www.exploration Plan_for_the_ISS.pdf.

[91] National Research Council, *Review of NASA Plans for the International Space Station*, 2006, http://www.nap.edu/catalog/11512.html.

[92] P.L. 109-155, Section 507.

[93] National Aeronautics and Space Administration, *NASA Report to Congress Regarding a Plan for the InternationalSpace Station National Laboratory*, May 2007, http://www.nasa.gov/pdf/181149main_ISS_ National_Lab_Final_Report_rev2.pdf.

[94] P.L. 110-422, Section 602.

[95] National Aeronautics and Space Administration, Congressional Budget Justification for FY2010, http://www.nasa.gov/pdf/345225main_FY_2010_UPDATED_final_5-11-09_with_cover.pdf, p. SPA-15.

[96] *Ibid.*

ISS Service Life Extension

The 2015 end date for U.S. utilization of the ISS arises from the engineering specifications of the U.S. ISS components, which were designed for a 15-year lifetime from the date of deployment. The various components were launched sequentially during the assembly process, but the nominal reference point is considered to be the launch of the U.S. laboratory module Destiny in February 2001. Despite the 15-year specification, past experience "clearly indicates that systems are capable of performing safely and effectively for well beyond their original design lifetime" if properly maintained, refurbished, and validated. The first milestones for a decision on service life extension will occur in 2014.[97]

Many ISS advocates want to continue utilization past 2015 in order to receive a greater return on the cost and effort that have been invested in ISS construction. The international partners issued a joint statement in July 2008 calling for operations to continue beyond 2015. Russia has stated that, if necessary, it will continue operations on its own. (Some analysts doubt that this would be technically feasible.) The Augustine committee found that extending ISS operations would "significantly enhance" the return on investment to both the United States and its international partners, while a decision not to extend operations would "significantly impair U.S. ability to develop and lead future international spaceflight partnerships." Three of the five options considered by the Augustine committee include the extension of ISS operations until 2020. Congress has directed NASA to ensure that the ISS remains viable through at least 2020 and to take no steps that would preclude continued U.S. utilization after 2015.[98]

In addition to cost, extending the life of the ISS would require overcoming several technical challenges. At present, failed parts are returned to Earth in the space shuttle for refurbishment. After the conclusion of the shuttle program, this repair strategy will likely no longer be possible. Most of the cargo vehicles that are being considered for the post-shuttle period are not capable of returning cargo back to Earth.[99] Instead, new parts would need to be manufactured and sent up, but even this may be impossible in a few cases, as some ISS parts are too large for any of the planned post-shuttle cargo alternatives. Last but not least, as ISS components reach the end of their 15-year design life, they will need to be recertified, which is a potentially complex and costly process.

Alternatives to service life extension also pose challenges. Some have suggested that the ISS could be operated by the other international partners with little or no U.S. participation. The Augustine committee found that this would be "nearly impossible" within the available budgets of the partner space agencies and because export controls would limit the direct support NASA could provide to foreign space agencies. Another option that has been proposed is to "mothball" the ISS for later use. The Augustine committee found that operating the ISS unoccupied would increase the risk of loss by a factor of five and also increase the risk of uncontrolled reentry into Earth's atmosphere, which would pose a hazard to people on the ground. Even deorbiting the ISS in a controlled manner is a challenging task. The Augustine committee found that no existing or currently planned vehicle is capable of

---

[97] National Aeronautics and Space Administration, *NASA Report to Congress Regarding a Plan for the International Space Station National Laboratory*, p. 5.

[98] P.L. 110-422, Section 601.

[99] After delivering their payloads, they are designed to burn up in the atmosphere or crash into the ocean.

this task. It projected that the cost of developing one, or of disassembling the ISS on-orbit and deorbiting the major components separately, could be $2 billion or more.[100]

Under the Administration's FY2011 budget, ISS operations would be extended to at least 2020.

Post-Shuttle Access to the ISS

The U.S. space shuttle has been the major vehicle taking crews and cargo to and from the ISS. Russian Soyuz spacecraft also carry both crews and cargo. Russian Progress spacecraft carry cargo only, as they are not designed to survive reentry into the Earth's atmosphere. A Soyuz is always attached to the station as a "lifeboat" in case of an emergency. The "lifeboat" Soyuz must be replaced every six months.

Unless the shuttle program is extended, paying Russia for flights on the Soyuz is the only short-term option for U.S. human access to the ISS. In 2009, in order to permit such payments, Congress extended a waiver of the Iran, North Korea, and Syria Nonproliferation Act (P.L. 106178 as amended) until July 1, 2016.[101]

One element of NASA's plans for ensuring cargo access to the ISS during the gap is the Commercial Orbital Transportation Services (COTS) program to develop commercial capabilities for cargo spaceflight. Under the COTS program, SpaceX Corporation is developing a vehicle known as Dragon, and Orbital Sciences Corporation is developing a vehicle known as Cygnus. Both would be cargo-only and would have about one-eighth the capacity of the space shuttle.[102] Only Dragon would be capable of returning cargo to Earth as well as launching it into space. Neither has yet flown into space. In the NASAAuthorization Act of 2008, Congress directed NASA to develop a contingency plan for post-shuttle cargo resupply of the ISS in case commercial cargo services are unavailable.[103] This plan was due to Congress by October 2009. It had not yet been completed as of December 2009.[104]

Noncommercial alternatives for cargo, in addition to the Russian Progress, include the European Automated Transfer Vehicle (ATV) and the Japanese H-II Transfer Vehicle (HTV). The first ATV was launched in March 2008 and carried out docking demonstrations with the ISS the following month. The first HTV was launched in September 2009 and also docked successfully with the ISS. Contracting with Russia for use of the Progress would probably require passing an additional waiver of the Iran, North Korea, and Syria Nonproliferation Act. Like the Dragon and Cygnus, the ATV, HTV, and Progress all have significantly smaller cargo capacity than the space shuttle.[105] None of the noncommercial alternatives is capable of returning cargo to Earth.

---

[100] Review of U.S. Human Spaceflight Plans Committee, *Seeking a Human Spaceflight Program Worthy of a GreatNation*, pp. 53-54.

[101] Consolidated Security, Disaster Assistance, and Continuing Appropriations Act of 2009 (P.L. 110-329), Section 125. For more information, see CRS Report RL34477, *Extending NASA's Exemption from the Iran, North Korea, and Syria Nonproliferation Act*, by Carl E. Behrens and Mary Beth Nikitin.

[102] Review of U.S. Human Spaceflight Plans Committee, *Seeking a Human Spaceflight Program Worthy of a GreatNation*, Fig. 4.2.2-1, p. 53.

[103] P.L. 110-422, Section 603.

[104] NASA Office of Legislative Affairs, personal communication, December 17, 2009.

[105] Review of U.S. Human Spaceflight Plans Committee, *Seeking a Human Spaceflight Program Worthy of a Great Nation*, Fig. 4.2.2-1, p. 53. The Progress has about 10% the cargo capacity of the shuttle. The HTV and ATV have about 20%.

FUTURE ACCESS TO SPACE

Whether or not the shuttle program is extended, and whatever option is chosen for access to the ISS during the gap, in the long term new vehicles will be needed to carry humans and cargo into space. Under current plans, these are the crew capsule Orion, the Ares I rocket to launch Orion into low Earth orbit, and the heavy-lift Ares V rocket to launch cargo. A variety of alternatives to these plans have been proposed, including reliance on commercial launch services.

Orion and Ares

One option is to continue with the status quo (i.e., Orion, Ares I, and eventually Ares V). Development of Orion and Ares I is well under way by NASA and its contractors. Development of Ares V has not begun, but Ares I and Ares V are to share some components. The first crewed flight of Orion and Ares is scheduled for 2015. The Augustine committee concluded that a 2017 date is more realistic and that a delay until 2019 is possible.

Orion is similar to an enlarged Apollo capsule. It is designed to carry six astronauts and to operate in space for up to six months. An upgraded version would be required for travel to the Moon or beyond. The Augustine committee concluded that Orion "will be acceptable for a wide variety of tasks in the human exploration of space" but expressed concern about its operational cost once developed. The committee suggested that a smaller, lighter, four-person version could reduce operations costs for support of the ISS by allowing landing on land rather than in the ocean and by enabling simplifications in the launch-abort system.[106]

The Ares I rocket is designed to be a high-reliability launcher that, when combined with Orion, will yield a crew transport system with an estimated 10-fold improvement in safety relative to the space shuttle. The development of Ares I has encountered some technical difficulties, but the Augustine committee characterized these as "not remarkable" and "resolvable." On the other hand, the committee concluded that the ultimate utility of Ares I has been diminished by schedule delays, as it will likely not be available until after the currently planned termination of the ISS.[107]

Ares V is designed to be capable of launching 160 metric tons of cargo into low Earth orbit.[108] (By comparison, the space shuttle has a cargo capacity for ISS resupply missions of about 16 metric tons, and the ISS, which was launched in pieces over a decade, weighs a total of 350 metric tons.[109]) For human missions beyond low Earth orbit, Ares V would launch equipment into orbit for rendezvous with an Orion launched by an Ares I. At present, Ares V is only a conceptual design. The Augustine committee described it as "an extremely capable

[106] Review of U.S. Human Spaceflight Plans Committee, *Seeking a Human Spaceflight Program Worthy of a Great Nation*, p. 58.

[107] Review of U.S. Human Spaceflight Plans Committee, *Seeking a Human Spaceflight Program Worthy of a GreatNation*, p.60.

[108] Review of U.S. Human Spaceflight Plans Committee, *Seeking a Human Spaceflight Program Worthy of a GreatNation*, p. 59.

[109] Review of U.S. Human Spaceflight Plans Committee, *Seeking a Human Spaceflight Program Worthy of a Great Nation*, pp. 53 and 64.

rocket" but estimated that under current budget plans, it is unlikely to be available until the late 2020s.

Government Alternatives to Ares V

The Augustine committee identified three categories of heavy-lift launchers that could be alternatives to Ares V: a scaled-down version of Ares V called Ares V Lite; a rocket derived from the design of the space shuttle; and a rocket derived from the Evolved Expendable Launch Vehicle. Unlike Ares V, each of these could be rated to carry humans (in an Orion capsule) as well as cargo.[110]

The Ares V Lite would be a slightly lower-performance version of the Ares V, capable of launching about 140 metric tons rather than 160. For human missions beyond Earth orbit, two launches of Ares V Lite, rather than one of Ares I and one of Ares V, would considerably increase the payload that could be carried to the destination. Some human missions beyond Earth orbit could be accomplished with a single Ares V Lite launch.[111]

Shuttle-derived vehicles would use the same main engines, solid rocket boosters, and external tanks as the space shuttle. They could be either in-line or side-mount. In other words, the payload could be mounted either on top or on the side. One example of the in-line option is the Jupiter design advocated by DIRECT, a group ostensibly led by NASA engineers working anonymously on their own time.[112] The Augustine committee did not compare the in-line and side-mount variants in detail, but it considered the side-mount option to be inherently less safe when carrying a crew. A shuttle-derived launcher would likely be able to lift 90 to 110 metric tons into orbit.

The Evolved Expendable Launch Vehicle program was a U.S. Air Force program that resulted in the development of the Delta IV and Atlas V rockets. Current EELV systems are not rated to carry humans. In testimony to the Augustine committee, the Aerospace Corporation stated that a human-rated variant of the Delta IV Heavy would be capable of carrying Orion to the ISS.[113] A super-heavy EELV variant could carry a cargo payload of about 75 metric tons. The Augustine committee concluded that using an EELV variant to launch Orion would only make sense if a super-heavy EELV variant were to be selected for heavy-lift cargo launch.

In addition to differences in capability, the Augustine committee found that these alternatives differ in their life-cycle costs, operational complexity, and "way of doing business." The committee concluded that Ares V Lite would be the most capable; that a shuttle derivative would take maximum advantage of existing infrastructure, facilities, and production capabilities; and that an EELV derivative could potentially have the lowest operating costs, but would require a significant restructuring of NASA. The committee noted

---

[110] NASA requires crewed space systems to be certified as human-rated. See *Human-Rating Requirements for SpaceSystems*, NASA Procedural Requirement 8705.2B, http://nodis3.gsfc.nasa.gov/displayDir.cfm?Internal_ID=N_PR_8705_002B_. Engineering requirements such as redundancy and fault tolerance are greater for human-ratedsystems than for uncrewed systems.

[111] Review of U.S. Human Spaceflight Plans Committee, *Seeking a Human Spaceflight Program Worthy of a Great Nation*, pp. 66-67.

[112] See http://www.directlauncher.com/.

[113] Review of U.S. Human Spaceflight Plans Committee, *Seeking a Human Spaceflight Program Worthy of a Great Nation*, p. 69.

that each alternative has strong advocates and that "the claimed cost, schedule, and performance parameters include varying degrees of aggressiveness."[114] It did not explicitly recommend any of the alternatives over the others.

Commercial Services as an Alternative to Ares I

Another alternative is to rely on the private sector to develop commercial crew transport services. The COTS program is already considering this possibility for ISS crew transfer and crew rescue. (This COTS capability is known as COTS D.) The NASA Authorization Act of 2008 directed NASA to make use of commercial crew services to the maximum extent practicable consistent with safety requirements.[115] The Augustine committee more broadly considered relying on commercial services in lieu of Ares I for all crew access to low Earth orbit. It included this approach in all the options it evaluated except Option 1 (status quo) and Option 3 (status quo with increased funding).

The Augustine committee concluded the following:[116]

- Considering that all U.S. crew launch systems to date have been built by industry, "there is little doubt" that the U.S. aerospace industry is capable of building and operating a four-passenger "crew taxi" to low Earth orbit.
- Because of the importance of crew safety, commercial crew transport services would need to include "a strong, independent mission assurance role for NASA."
- If the service were developed so as to meet commercial needs as well as NASA's, there would be private-sector customers to share operating costs with NASA. In that case, the cost of the program to NASA would be about $5 billion, and a service could be in place by 2016.
- If the private sector effort were to fail in mid-program, the task of crew transport would revert to NASA. NASA should continue development of Orion and move quickly toward the development of a human-ratable heavy-lift rocket as a fallback option to mitigate this risk.

The Administration's FY2011 budget proposals include reliance on commercial crew transport services. Unlike the Augustine committee's proposal, the Administration budget would not support continued development of Orion except for emergency escape from the International Space Station.

The Augustine committee found that the commercial space industry is "burgeoning," and concluded that creating an assured initial market would eventually have the potential—"not without risk"—to significantly reduce costs to the government.

On the other hand, the Augustine committee also pointed out that developing Ares I would give the NASA workforce, which has not developed a new rocket in more than 20

---

[114] Review of U.S. Human Spaceflight Plans Committee, *Seeking a Human Spaceflight Program Worthy of a Great Nation*, p. 72.

[115] P.L. 110-422, Section 902.

[116] Review of U.S. Human Spaceflight Plans Committee, *Seeking a Human Spaceflight Program Worthy of a Great Nation*, pp. 70-72.

years, an opportunity to gain experience with the simpler Ares I system before beginning development of the more complex Ares V.[117]

Congressional policy makers may wish to consider that instead of contracting with the private sector for crew services, NASA could continue to contract for the use of Russian Soyuz vehicles. This would probably require Congress to further extend its waiver of the Iran, North Korea, and Syria Nonproliferation Act. The Augustine committee concluded that while reliance on Soyuz on an interim basis is acceptable, longer-term use would not be. It argued that "an important part of sustained U.S. leadership in space is the operation of our own domestic crew launch capability."[118]

Issues for Congress

Cost, safety, capability, and technical feasibility are key issues for Congress in considering these alternatives. The Augustine committee used minimum criteria for safety, capability, and technical feasibility as filters to eliminate unacceptable alternatives. It gave little information that would distinguish clearly between the remaining, acceptable alternatives on the basis of these factors. It concluded that of the acceptable alternatives, all would require roughly the same additional funding: about $3 billion per year above current plans. It noted that NASA has a history of designing for maximum capability at minimum development cost, with less focus on operational cost, and it suggested that NASA programs would be more sustainable if designed for minimum life-cycle cost.[119]

Safety

Space travel is inherently dangerous. Nevertheless, NASA's policy is that "safety is and will always be our number one priority in everything we do."[120] The Augustine committee described safety as a *sine qua non.*[121] Analysts and policy makers generally agree with this emphasis, but some have concerns about whether it is matched by NASA's implementation of its safety policies and procedures.

The Columbia Accident Investigation Board found in 2003 that "throughout its history, NASA has consistently struggled to achieve viable safety programs and adjust them to the constraints and vagaries of changing budgets.... NASA's safety system has fallen short of the mark."[122] It concluded that "a broken safety culture," including a "reliance on past success as a substitute for sound engineering practices," was an organizational cause of the *Columbia*

[117] Review of U.S. Human Spaceflight Plans Committee, *Seeking a Human Spaceflight Program Worthy of a Great Nation*, p. 69.

[118] Review of U.S. Human Spaceflight Plans Committee, *Seeking a Human Spaceflight Program Worthy of a Great Nation*, p. 69.

[119] Review of U.S. Human Spaceflight Plans Committee, *Seeking a Human Spaceflight Program Worthy of a Great Nation*, p. 68.

[120] Jeffrey Hanley, Manager, NASA Constellation Program, testimony before the House Committee on Science and Technology, Subcommittee on Space and Aeronautics, December 2, 2009.

[121] Review of U.S. Human Spaceflight Plans Committee, *Seeking a Human Spaceflight Program Worthy of a GreatNation*, p. 9.

[122] Columbia Accident Investigation Board, vol. 1, p. 192.

disaster.[123] It found that one contributing factor was "intense schedule pressure," which had also been identified as an organizational cause of the space shuttle *Challenger* disaster in 1986.[124] It recommended that NASA establish a technical engineering authority, reporting directly to the NASAAdministrator rather than to the space shuttle program, that independently verifies launch readiness and has sole authority to grant waivers for technical standards.[125] In response to these findings, NASA has made many changes, including the establishment of an independent NASA Engineering and Safety Center under the auspices of the headquarters Office of Safety and Mission Assurance.

Nevertheless, some analysts see signs that potential problems remain. The deadline of 2010 to complete construction of the space station and stop flying the space shuttle created schedule pressure for both programs until NASA converted it from a hard deadline to a flexible goal. In 2006, NASA decided to a proceed with a shuttle mission, even though the Chief Engineer and the head of the Office of Safety and Mission Assurance recommended against the launch because of an issue with the shuttle ice-frost ramps that they characterized as "probable/catastrophic."[126] Some observers saw signs of "reliance on past success" in NASA's justification for this decision: the NASAAdministrator disagreed with the "probable" characterization because "we have 113 flights with this vehicle ... and while we've had two loss of vehicle incidents, they've not been due to ice-frost ramps."[127] (The two officials who recommended against launch stated that they were comfortable with the decision to overrule them because "the risk was to the vehicle and not the crew.")[128] A member of NASA's Aerospace Safety Advisory Panel testified in late 2009 that describing safety as a *sine qua non* "oversimplifies a complex and challenging problem" and that NASA "has given serious consideration only recently" to the establishment of safety requirements for commercial crew transport services.[129]

NASA argues that it continues to implement initiatives to improve safety. These include greater emphasis on training and qualification of safety professionals; an emphasis on "safety culture," including more open communication and clear appeal paths to the Administrator for safety-related dissenting opinions; more modeling and validation of software requirements; and improved tools for knowledge and requirements management.[130] The design process for Orion and Ares I is "riskinformed," including the systematic identification and elimination of hazards and the mitigation of remaining risks via effective abort systems.[131]

[123] Columbia Accident Investigation Board, vol. 1, pp. 184 and 9.

[124] Columbia Accident Investigation Board, vol. 1, p. 97.

[125] Columbia Accident Investigation Board, vol. 1, p. 193.

[126] "July 1 Shuttle Launch OK'd with Some Reservations," *Aerospace Daily*, June 20, 2006.

[127] Ibid.

[128] "Shuttle Launch Holdouts Explain No-Go Recommendations," *Aerospace Daily*, June 22, 2006.

[129] John Marshall, NASA Aerospace Safety Advisory Panel, testimony before the House Committee on Science and Technology, Subcommittee on Space and Aeronautics, December 2, 2009, http://democrats.

[130] Bryan O'Connor, Chief, NASA Office of Safety and Mission Assurance, testimony before the House Committee on Science and Technology, Subcommittee on Space and Aeronautics, December 2, 2009, http://democrats.

[131] Joseph R. Fragola, Valador Inc., testimony before the House Committee on Science and Technology, Subcommitteeon Space and Aeronautics, December 2, 2009, http://democrats Commdocs/hearings/ 2009/Space/2dec/Fragola_Testimony.pdf. (Dr. Fragola was a member of the ESAS.)

Other Issues

Some experts assert that none of the options under consideration would be capable of conducting on-orbit repair missions, as the space shuttle is.[132] They note that the shuttle has made several missions to repair and upgrade the Hubble space telescope.[133] The Hubble telescope is unique, however, in being designed to be serviced by astronauts. Most satellites are never serviced, and if there is only an occasional need for such a capability, it may be cheaper simply to replace them if they fail, rather than invest in a rarely used servicing capability.

NASA has a history of new vehicle development programs ending before their completion. The National Aerospace Plane, intended as a replacement for the space shuttle, was announced in 1986 and cancelled in 1992.[134] The X-33 and X-34 projects, intended to demonstrate technology for a commercial space shuttle replacement, were announced in 1996 and cancelled in 2001.[135]

Congress may wish to consider whether there is an inherent risk in abandoning the agency's current plans even if another alternative seems preferable on its merits. The potential for this risk may be reduced for the heavy-lift alternatives currently being considered because they are all closely related either to the currently planned systems (Ares V Lite to Ares I and Ares V) or to existing operational systems (shuttle-derived rockets to the space shuttle and EELV-derived rockets to the Delta IV or Atlas V).

Destinations for Human Exploration

In considering possible modifications to the Vision, space policy experts and other interested observers have suggested various alternative goals. For example, some have proposed that Mars should be the immediate objective, rather than returning to the Moon first. Others have suggested human missions to asteroids or other solar system destinations.

The most prominently discussed alternative, especially before the Augustine committee released its report, is to proceed to Mars directly.[136] The Augustine committee rejected this possibility because it considered current technology insufficiently developed to make a Mars mission safe. It found that Mars is "unquestionably the most interesting destination in the inner solar system" and the "ultimate destination for human exploration" but "not the best first destination."[137]

[132] For example, see Joel Achenbach, "NASA Scientist Decries Agency's Plans," *Washington Post*, May 23, 2009.
[133] See CRS Report RS21767, *Hubble Space Telescope: NASA's Plans for a Servicing Mission*, by Daniel Morgan.
[134] Columbia Accident Investigation Board, vol. 1, pp. 110-111.
[135] Columbia Accident Investigation Board, vol. 1, p. 111.
[136] See, for example, the advocacy of the Mars Society, http://www.marssociety.org/.
[137] Review of U.S. Human Spaceflight Plans Committee, *Seeking a Human Spaceflight Program Worthy of a Great Nation*, p. 21.

Destination	Public Engagement	Science	Human Research	Exploration Preparation
Lunar Flyby/Orbit	Return to Moon, "any time we want"	Demo of human robotic operation	10 days beyond radiation belts	Beyond LEO shakedown
Earth Moon L1	"On-ramp to the inter-planetary highway"	Ability to service Earth Sun L2 spacecraft at Earth Moon L1	21 days beyond the belts	Operations at potential fuel depot
Earth Sun L2	First human in "deep space" or "Earth escape"	Ability to service Earth Sun L2 spacecraft at Earth Sun L2	32 days beyond the belts	Potential servicing, test airlock
Earth Sun L1	First human "in the solar wind"	Potential for Earth/Sun science	90 days beyond the belts	Potential servicing, test in-space habitation
NEO's	"Helping protect the planet"	Geophysics, Astrobiology, Sample return	150-220 day, similar to Mars transit	Encounters with small bodies, sample handling, resource utilization
Mars Flyby	First human "to Mars"	Human robotic operations, sample return?	440 days, similar to Mars out and return	Robotic operations, test of planetary cycler concepts
Mars Orbit	Humans "working at Mars and touching bits of Mars"	Mars surface sample return	780 days, full trip to Mars	Joint robotic/human exploration and surface operations, sample testing.
Mars Moons	Humans "landing on another moon"	Mars moons' sample return	780 days, full rehearsal Mars exploration	Joint robotic/human surface and small body exploration

Source: Review of U.S. Human Spaceflight Plans Committee, *Seeking a Human Spaceflight Program Worthy of a Great Nation*, p. 41.

Notes: L1 and L2 refer to particular Lagrange points (see text). A NEO is a near-Earth object such as an asteroid or a comet.

Figure 1. Potential Activities at Alternative Human Exploration Destinations (in Addition to the Moon and Mars) as Evaluated by the Augustine Committee

A spacecraft that lands on either the Moon or Mars must overcome the lunar or Martian "gravity well" before returning to Earth.[138] The fuel required to accomplish this makes either destination challenging. As potential alternatives, the Augustine committee considered fly-by missions to either the Moon or Mars, missions that would orbit either the Moon or Mars, missions that would land on the moons of Mars, and missions to near-Earth objects such as asteroids or comets. They also considered missions to various Lagrange points. Lagrange points are special locations in space, defined relative to the orbit of the Moon around the Earth or the Earth around the Sun. They are planned locations for future unmanned science spacecraft, and some scientists believe they will be important in determining routes for future interplanetary travel. Possible activities at each of these destinations are shown in **Figure 1**.

Under the Administration's proposals, as articulated in the President's speech in April 2010, the first destination would be an asteroid, followed by an orbit of Mars and subsequently a Mars landing.

[138] At least one observer has suggested the launching humans to Mars without plans to return them to Earth. See Lawrence M. Krauss, "A One-Way Ticket to Mars," *New York Times*, September 1, 2009. It seems unlikely that this option would be politically viable.

ALTERNATIVES TO HUMAN EXPLORATION

Given the costs and risks of human space exploration, Congress could decide to curtail or postpone future human exploration missions and shift the emphasis of the nation's space program to other endeavors. The cost of human exploration is substantial, and according to the Augustine committee, it is not a continuum: there is an "entry cost" below which a successful program cannot be conducted at all.[139] Congress could decide that this minimum cost is not affordable. Similarly, no matter how energetically NASA addresses safety concerns, human spaceflight is an inherently risky endeavor. Congress could decide that the potential benefits are insufficient to justify the safety risks.

Several options are available as alternatives to human space exploration. Congress could seek to accomplish some of the same goals through other means, such as through robotic exploration. It could focus on technology development, in the hope of developing new technology that makes human spaceflight safer and more affordable in the future. It could focus on NASA's other activities, such as Earth science and aeronautics. Given sufficient funding, of course, all these options are also available in conjunction with human exploration rather than as alternatives to it. For example, the Augustine committee acknowledged that robotic exploration is important as a precursor to human exploration.

Robotic Exploration

Advocates of robotic missions assert that robotic exploration can accomplish outstanding science and inspire the public just as effectively as human exploration. The Mars rovers are a familiar example of a successful robotic science mission that has captured considerable public attention. Advocates also claim that robotic missions can accomplish their goals at less cost and with greater safety than human missions. They do not need to incorporate systems for human life support or human radiation protection, they do not usually need to return to Earth, and they pose no risk of death or injury to astronauts. Some analysts assert further that exploring with humans "rules out destinations beyond Mars."[140] Given that current plans include no destinations beyond Mars and treat Mars itself only as a long-term goal, this last limitation may not be important in the near term, even if it is correct.

Advocates of human missions note that science is not NASA's only purpose and claim that human exploration is more effective than robotic exploration at such intangible goals as inspiring the public, enhancing national prestige, and satisfying the human urge to explore and discover. They assert that even considering science alone, human missions can be more flexible in the event of an unforeseen scientific opportunity or an unexpected change in plans. As support for this assertion, they often cite the human missions to repair and upgrade the Hubble telescope. On the other hand, the Hubble repairs and upgrades required extensive planning and the development of new equipment. They were not a real-time response to an unexpected event. Moreover, robotic missions can also sometimes be modified to respond to

---

[139] Review of U.S. Human Spaceflight Plans Committee, *Seeking a Human Spaceflight Program Worthy of a Great Nation*, p. 23.

[140] See, for example, Robert L. Park, *What's New*, September 11, 2009, http://bobpark.physics.umd.edu/WN09/wn091109.html.

opportunities and mishaps, through software updates and other changes worked out by scientists and engineers back on Earth.

A few analysts portray robotic exploration as an alternative to human exploration. For the most part, however, the two alternatives are considered complementary, rather than exclusive. The Augustine committee, for example, found that without both human and robotic missions, "any space program would be hollow."[141] In addition, many analysts consider that in the absence of human missions, support for NASA as a whole would dwindle, and fewer resources would be available for robotic missions as well.

Emphasize Technology Development

If congressional policy makers were to conclude that cost and safety concerns make a human exploration program unaffordable or undesirable in the near term, they might seek to scale back NASA's human spaceflight program and focus on technology development, in the hope that improved technology will make the costs and risks of space travel more attractive in the future.

The strategy of developing improved technology and acquiring greater expertise could take many forms. It could complement a continuing, aggressive program of human exploration. For example, it is similar, in some ways, to the Augustine committee's suggestion (in its "flexible path" option) of visiting a series of less challenging destinations before attempting a Moon landing. It could also accompany a program of human missions in low Earth orbit without immediate plans for more distant destinations. It could even be part of a program that abandoned human spaceflight in the near term. Developing new technology effectively would likely be difficult, however, without a means of testing it in realistic missions. A program without any human spaceflight at all would risk losing existing expertise through inactivity.

The Augustine committee, the National Academies, and the Administration's FY2011 budget proposals all recommended a greater emphasis on technology development as a complement to an ongoing program of human spaceflight. The Augustine committee described NASA's space technology program as "an important effort that has significantly atrophied over the years."[142] It recommended that technology development be closely coordinated with ongoing programs, but conducted independently of them. The National Academies also recommended that NASA revitalize its technology development program. Like the Augustine committee, the Academies concluded that this program should be conducted independently. They recommended that NASA establish a DARPA-like technology development organization that reports directly to the Administrator.[143]

---

[141] Review of U.S. Human Spaceflight Plans Committee, *Seeking a Human Spaceflight Program Worthy of a Great Nation*, p. 114.

[142] Review of U.S. Human Spaceflight Plans Committee, *Seeking a Human Spaceflight Program Worthy of a Great Nation*, p. 112.

[143] National Research Council, *America's Future in Space*, pp. 61-62. DARPA is the Defense Advanced Research Projects Agency, http://www.darpa.mil/, a frequent model for technology development agencies within other departments.

OTHER SPACE POLICY ISSUES

In addition to the programmatic and prioritization issues that are the main focus of this chapter, NASA faces some cross-cutting challenges, such as acquisition and financial management, and issues involving its relationships with other agencies and the commercial space launch industry.

NASA Acquisition and Financial Management

Since 1990, the GAO has identified acquisition management at NASA as a high-risk area for the federal government.[144] Although a 2009 update noted that "NASA has made a concerted effort to improve its acquisition management," it also stated that "since fiscal year 2006, 10 out of 12 ... major development projects in implementation exceeded their baseline thresholds."[145] NASA has issued an improvement plan in response to GAO's finding.[146] In the NASAAuthorization Act of 2005, Congress established requirements for baselines and cost controls.[147] These requirements include additional reviews of any program that appears likely to exceed its baseline cost estimate by 15% or its baseline schedule by six months and a prohibition on continuing any program that exceeds its cost baseline by 30% unless Congress specifically authorizes the program to continue. These requirements are similar to the Nunn-McCurdy cost-containment requirements for the Department of Defense.

In November 2008, the NASA Inspector General identified financial management as one of NASA's most serious challenges.[148] The Inspector General's report found continuing weaknesses in NASA's financial management processes and systems, including its internal controls over property accounting. It noted that these deficiencies have resulted in disclaimed audits of NASA's financial statements since FY2003, largely because of data integrity issues and a lack of effective internal control procedures. According to the report, NASA has made progress in addressing these deficiencies, but the FY2008 audit of its financial statements showed that deficiencies still exist. The FY2009 audit was again disclaimed.[149]

In the conference report on the Consolidated Appropriations Act, 2010 (H.Rept. 111-366), Congress expressed its continuing concern about NASA's

> pattern of lax fiscal management and oversight,ranging from the administration ofenhanced use lease receipts, insufficient evidentiary support for amounts in NASA's budget

---

[144] Government Accountability Office, *High-Risk Series: An Update*, GAO-09-271, January 2009. As of this 2009 update, GAO listed 27 high-risk areas.

[145] *Ibid.*, p. 77.

[146] National Aeronautics and Space Administration, *NASA Plan for Improvement in the GAO High-Risk Area of Contract Management*, October 31, 2007, updated January 31, 2008, http://www.nasa.gov/pdf/ 270426main_NASA_High-RiskCAP-Jan2008Final.pdf.

[147] P.L. 110-422, Section 103.

[148] National Aeronautics and Space Administration, Office of the Inspector General, *NASA's Most Serious Management and Performance Challenges*, November 10, 2008, http://oig.nasa.gov/NASA2008Management Challenges.pdf.

[149] See House Committee on Science and Technology, Subcommittee on Investigations and Oversight and Subcommittee on Space and Aeronautics, *Independent Audit of the National Aeronautics and Space Administration*, hearing December 3, 2009.

execution data, improper execution of its authority in the construction program, and increasing numbers of contract awards being protested.

U.S. Space Policy Governance

A variety of governmental and nongovernmental organizations help to coordinate and guide U.S. space policy. These include the Office of Science and Technology Policy (OSTP) and the National Science and Technology Council (NSTC), both in the Executive Office of the President, as well as outside advisory groups, such as the NASAAdvisory Council,[150] committees of the National Academies,[151] and independent committees such as the Augustine committee.

The National Academies have recommended that the President

> task senior executive-branch officials to align agency and department strategies; identify gaps or shortfalls in policy coverage, policy implementation, and resource allocation; and identify new opportunities for space-based endeavors that will help to address the goals of both the U.S. civil and national security space programs.[152]

The Obama Administration has stated that it intends to reestablish the National Aeronautics and Space Council (NASC), "which will report to the President and oversee and coordinate civilian, military, commercial and national security space activities."[153] The NASC was established along with NASA itself by the National Aeronautics and Space Act of 1958 (P.L. 85-568). It was most active during the Kennedy Administration, when it recommended, among other policies, the Apollo program to send humans to the Moon. Some analysts attribute its influence during this period to the fact that it was chaired by Vice President Johnson. The NASC was abolished in 1973, reestablished in 1989 as the National Space Council, then abolished again in 1993, with its functions absorbed into the NSTC.

Some aspects of space policy are documented in a formal presidential statement of national space policy. In 2006, the Bush Administration issued such a statement,[154] replacing a previous one that had been in place for 10 years.[155] The 2006 policy established principles and goals for U.S. civilian and national security space programs and set guidelines for a few specific issues such as the use of nuclear power in space and the hazard of debris in orbit. It defined the space-related roles, responsibilities, and relationships of NASA and other federal agencies, such as the Department of Defense and the Department of Commerce.

---

[150] *NASA Advisory Council (NAC)*, http://www.nasa.gov/offices/nac/home/index.html.

[151] Such as the Space Studies Board, http://sites.nationalacademies.org/SSB/index.htm, the Aeronautics and Space Engineering Board, http://sites.nationalacademies.org/deps/ASEB/, and their committees and subcommittees.

[152] National Research Council, *America's Future in Space*, p. 8.

[153] Office of Science and Technology Policy, *Issues: Technology*, http://www.ostp.gov/cs/issues/technology.

[154] National Security Presidential Directive (NSPD) 49, *U.S. National Space Policy*, August 31, 2006,http://www.fas.org/irp/offdocs/nspd/space.html.

[155] Presidential Decision Directive NSC-49/NSTC-8, *National Space Policy*, September 14, 1996, http://www .fas.org/spp/military/docops/national/nstc-8.htm.

U.S. National Security Space Programs

National security space programs, conducted by the Department of Defense (DOD) and the intelligence community, are less visible than NASA, but their budgets are comparable to NASA's. A key issue for them is how to avoid the cost growth and schedule delays that have characterized several recent projects. A recent analysis of numerous GAO reports directly related to DOD space programs found 16 recommendations still categorized as "open," including topics such as space acquisitions, polar-orbiting environmental satellites, DOD's operationally responsive space concept, and training of Air Force space personnel.[156] Further discussion of national security space programs is beyond the scope of this chapter.

NASA's Relationship with NOAA

Congressional policy makers have taken long-standing interest in NASA's relationship with the National Oceanic and Atmospheric Administration (NOAA), which operates Earth observing satellites for weather forecasting and other purposes. The NASAAuthorization Act of 2005 mandated the establishment of a joint NASA-NOAA working group; required NASA and NOAA to submit a joint annual report on coordination each February; directed NASA and NOAA to evaluate NASA science missions for their operational capabilities and prepare transition plans for those with operational potential; and directed NASA not to transfer any Earth science mission or Earth observing system to NOAA until a transition plan has been approved and funds have been included in the NOAA budget request.[157] In the NASA Authorization Act of 2008, Congress directed OSTP to develop a process for transitions of experimental Earth science and space weather NASA missions to operational status under NOAA, including the coordination of agency budget requests;[158] mandated a National Academies study of the governance structure for U.S. Earth observation programs at NASA, NOAA, and other agencies, to be transmitted to Congress by April 2010;[159] and mandated a National Academies assessment of impediments to interagency cooperation on space and Earth science missions, to be submitted to Congress by January 2010.[160]

The U.S. Commercial Space Industry

Industry has long had an important role in both space launch and the development and operation of commercial satellites. Although the commercial satellite launch business has dropped off in recent years, many analysts expect the industry to expand as space tourism develops and NASA begins to rely more on the commercial sector for space transportation.

---

[156] Marcia Smith, "DOD Has Mixed Record on Implementing GAO Space Program Recommendations," SpacePolicyOnline.com, December 28, 2009. For an explanation of the operationally responsive space concept, see National Security Space Office, "Operationally Responsive Space," http://www .acq.osd.mil/nsso/ors/ors.htm.

[157] P.L. 109-155, Section 306.

[158] P.L. 110-422, Section 204.

[159] P.L. 110-422, Section 202.

[160] P.L. 110-422, Section 507.

The prospect of space tourism on commercial vehicles is becoming increasingly likely. With the exception of five suborbital demonstration flights in 2004, private space travel has until now been limited to the purchase of trips to the International Space Station on Russian Soyuz spacecraft. A number of commercial companies are now developing reusable spacecraft to carry private individuals on short-duration flights into the lower reaches of space. Concurrently, several companies and states are developing spaceports to accommodate anticipated increases in commercial space launches. The safety of commercial space launches, spaceports, and space tourism are regulated by the Federal Aviation Administration (FAA). According to the GAO, the FAA faces a number of challenges in commercial space regulation, including maintaining sufficient space expertise to conduct proper oversight, avoiding conflicts between its regulatory and promotional roles, and integrating spacecraft into the air traffic control system.[161]

Export control regulations administered by the Department of State under the International Traffic in Arms Regulations (ITAR) have often been a concern for this industry. The regulations limit the export of satellites and related components because of the potential for their use in military systems. In order to expand opportunities for U.S. industry, some analysts and policy makers have advocated transferring the regulation of these technologies from ITAR to the Export Administration Regulations administered by the Department of Commerce.[162]

The development of commercial vehicles for cargo flights to the space station, and possibly to provide NASA with crew launch services into low Earth orbit, is discussed elsewhere in this chapter.

LEGISLATION IN THE 111TH CONGRESS

Appropriations for NASA are provided in the Commerce, Justice, Science (CJS) appropriations bill. For more information, see CRS Report R41161, *Commerce, Justice, Science, and Related Agencies: FY2011 Appropriations*.

For FY2010, the CJS bill was passed by the House and Senate as H.R. 2847. For final passage, it was included in the Consolidated Appropriations Act, 2010 (P.L. 111-117). For more information about FY2010 NASA appropriations, see CRS Report R40644, *Commerce, Justice, Science, and Related Agencies: FY2010 Appropriations*.

For FY2009 NASA appropriations legislation during the 111th Congress, including passage of the American Recovery and Reinvestment Act of 2009 (P.L. 111-5) and the Omnibus Appropriations Act, 2009 (P.L. 111-8), see CRS Report RL34540, *Commerce, Justice, Science and Related Agencies: FY2009 Appropriations*.

---

[161] Government Accountability Office, *Commercial Space Transportation: Development of the Commercial Space Launch Industry Presents Safety Oversight Challenges for FAA and Raises Issues Affecting Federal Roles*, GAO-10286T, December 2, 2009, http://www.gao.gov/new.items/d10286t.pdf.

[162] For more on this issue, see CRS Report RL31832, *The Export Administration Act: Evolution, Provisions, and Debate*, by Ian F. Fergusson; and Department of Commerce and Federal Aviation Administration, *Introduction to U.S. Export Controls for the Commercial Space Industry*, October 2008, http://www.faa.gov/about/office_org/headquarters_offices/ast/media/Intro%20to%20US%20Export%20Controls.pdf.

As noted at the beginning of this chapter, Congress is widely expected to act on a NASA authorization bill during the 111th Congress, but no such bill has yet been introduced.

Other space-related legislation in the 111th Congress includes the American Space Access Act (H.R. 1962) to extend the space shuttle program to 2015; a bill to extend the current third-party liability indemnification for commercial launch services companies by three years (H.R. 3819); and the Commercial Space Transportation Cooperative Research and Development Centers of Excellence Act of 2009 (H.R. 3853) to authorize grants to university consortia to establish centers in conjunction with NASA and industry.

SUMMARY OF MAJOR ISSUES FOR CONGRESS

In conclusion, the major space policy issues facing Congress include the following:

- Is there a national consensus for human exploration beyond Earth orbit, despitethe inherent risks and the substantial cost?
- If so, what destination or destinations should NASA's human exploration program explore? Should the Moon remain the target, as under current plans?Should there be a graduated sequence of targets as in the Augustine committee's "flexible path" option?
- If human exploration beyond Earth orbit is too costly or too dangerous, shouldNASA focus its efforts on human missions in Earth orbit, robotic exploration,technology development, other activities such as science and aeronautics, orsome combination of these?
- Should the space shuttle program be terminated at the end of 2010 (or in early2011) as currently planned? If so, how should the transition of the shuttleworkforce and facilities be managed? If the shuttle program is to be extended,what actions are needed to ensure the safety of its crews after 2010, and whatimpact will its continuing cost have on the availability of funds for other NASAprograms?
- Should U.S. use of the International Space Station continue past its currently planned termination at the end of 2015? If so, what impact will the additional cost have on the availability of funds for other NASA programs? If not, when and how should the International Space Station be deorbited?
- Are the currently planned Orion and Ares vehicles the best choices for delivering astronauts and cargo into space? Could commercial services take the place of Orion and Ares I for launching humans into low Earth orbit? If so, what steps should NASA take to develop that capability? Is Ares V the best heavy-lift launch option? If not, which of the alternatives is preferable?
- How should NASA's multiple objectives be prioritized? What is the proper balance between human spaceflight, science, aeronautics, and education programs, and how can the balance be maintained if the cost of the larger, more prominent programs grows?

In: Space Policy and Its Ramifications
Editor: John P. Ramos
ISBN: 978-1-61761-555-9

Chapter 3

FEASIBILITY AND IMPLICATIONS OF SELECT SPACE SECURITY TECHNOLOGY PROPOSALS OF THE FIRST DECADE OF THE 21ST CENTURY

Lars Rose*

University of British Columbia, Department of Materials Engineering,
Vancouver, BC, Canada

ABSTRACT

This chapter analyzes the feasibility of research and development related to select space security technology development plans. The proposals are presented in an international context where certain interest groups in the U.S.A. are not likely to be alone in having aspirations to space security technology. China, Russia, and France, among others, are just as likely to develop novel space security technology. Since the U.S. programs have received more attention and were arguably in more advanced stages of planning, if not development, this chapter focuses on U.S. national security space management and organization proposals, investigating five key technology projects that have been proposed and partially pursued. These systems are described in relation to the technical issues, engineering difficulties and physical limits involved. If appropriate, their implications are linked to the political impact of implementing these plans to technological maturity. Various different countries could conceivably answer the perceived threat inherent in the development of space security technology by emulating or outpacing each other in space militarization. While the U.S.A. arguably has more advanced security technology expertise, it is shown that other countries have already deployed some security technologies in space applications in the past, with less media attention. And, ultimately, the nations with the most space assets also have the most to lose from space militarization.

Keywords: Space policy, technology feasibility, research and development

* The information presented in this chapter does not necessarily reflect the views of the author's host institutes.

INTRODUCTION

Currently, no international policy for long-term political stability in outer space exists [for example: Goh 2004, Moltz 2007, Finarelli 2006, Stadd 2004, Kingwell 1990, Deblois 1999 and 2004, Jasani 1985 and 2001, Kries 2002, ISB 2004, Arcus 2003, Cohen 2000, USOTA 1995]. During the early years of the 21st century, the desire among some U.S. military personnel and politicians to develop space weapons received increased attention [Garwin 2005, Andrews 2001, Weiner 2005, Kleiman 2006, Hardesty 2005]. In some publications, space was labeled as a frontier that *has to* be protected through militarization [Andrews 2001, Zielinski 1996, Henry 2008, Hays 2002]. The notion derives from the constantly increasing reliance of many military and civil operations on space-based technology such as the Global Positioning System (GPS). Such heavy reliance on space technology creates vulnerability and countries with significant space assets thus desire to defend their assets against aggression, if necessary by force. As mentioned later in this chapter, a network-like constellation of micro-satellites could avoid the danger of system failure significantly, without the need to defend satellites that, for the moment, are often standalone models with almost no backup or redundancy.

Another factor driving the momentum for the development of space security technology is the global failure to develop a reliable ground-based missile defense [Weiner 2005, BBC 2001, Finney 2005]. The U.S.A. spent over $100 billion on R&D for missile defense during the past four decades, creating a significant sink for U.S. tax money; money that, for example, could have significantly boosted the development of novel, cleaner energy systems and investments to prevent or soften the economic downturn at the end of the first decade of the 21st century [Kammen 2006, Rose 2010, Dougherty 2009, Chen 2007]. Systems developed so far can still not reliably detect an international threat and react accordingly in due time, and it is unlikely that further spending will change that. Moreover, the installation of missile shields might result in other countries increasing the numbers of missiles they deploy and improving their countermeasure capabilities [Cirincione 2000]. However, the U.S.A. appears willing to evaluate the development of space-based interceptors as part of a global missile defense system [Collard-Wexler 2006].

How much, then, was spent on the U.S. military space technology ideas? Of the $440 billion U.S. defense budget in 2006 (excluding spending on wars in the Middle East), roughly $70 billion was spent on R&D activities [NP 2005, Hitchens 2006]. However, this money represents the overall official research budget of the U.S. military. Some $20 billion were spent annually under the general header of space defense; of this only a tiny amount, estimated at around $435 million, was designated annually for space security technology development [Lewis 2006, Beljac 2008]. With only a fraction of this allocated to specific research projects, it remains questionable whether the proposed systems can be developed to a mature usefulness. As will be discussed in this chapter, it is doubtful that even significant increases in funding could provide functioning systems in some of the outlined areas of research. Technology is, after all, only useful in terms of their original purpose if it works properly. Thus, the physical limitations of proposed future systems have to be taken seriously if unrealizable military research plans are to be avoided.

The development of space weapons is not necessarily solely a means of furthering some specific countries' interests. With human lives increasingly depending on technology in space, its market value increases. While this should be a strong argument against even considering any weapons technology, human nature has a strong urge to protect its assets. Consequently, all nations with significant assets in space strive to protect theirs. Existing and, even more so, rumored research programs in countries such as China, Russia, and the U.S.A. only serve to spin our society towards actually developing what currently may be no more than technological testbeds or proposed, but mostly untested, programs. The general perception propagated by international media that U.S. technology is far more advanced than that of other countries' technologies might be true but cannot be taken for granted, especially since media in the East (for example China and Russia) to some degree appear to be under state control, and not fully independent. The U.S.A. can certainly not causally be declared a sole aggressor in space, but it has done little to alleviate the international threat cycle. Since the U.S.A.'s plans are somewhat easier to follow than space technology research plans of other nations, this chapter focuses on its technologies, using them as a technological assessment for similar types that could potentially be developed globally.

There are many different proposals for space security technology systems. Simpler systems include those that cause actual electric disruption, by way of methods such as malicious programming or electromagnetic pulses. Such pulses are easily created and highly effective by igniting nuclear explosions in space, but such attacks are unspecific in their targets, invite nuclear retaliation, and are highly unlikely. This chapter will not include a discussion of such systems, but instead focus on specific hardware projects proposed by the U.S.A. within the past decades. Technological feasibility was often of less concern when novel creative ideas were brought forward. It is, however, this feasibility that should be addressed to avoid wasting tax revenues on systems that will never work satisfactorily because of inherent engineering obstacles and physical limits. It should be noted that some of the projects examined have already been discontinued, or submerged into clandestine operations (however unlikely), either as a result of public repercussions or because they have been deemed unrealizable wishful thinking [Oberg 2006 and 2007, Cornwell 2005, Gordon 1992, Sadeh 2009].

The U.S. administration regularly asserts the need to maintain its critical access to and use of security space assets, for example through the Offensive Counterspace Program. In this respect, the U.S.A. is no different than any other spacefaring nation. However, official texts usually avoid mentioning offensive space weaponry, since the priority of any nation with significant dependence on space technology is Space Situational Awareness [Tirpak 2004]. In addition, the U.S.A. has focused on reconstitution of space capabilities through so-called operationally responsive space systems as a response to space threats, rather than through offensive space control [Sega 2007, Cebrowski 2004 and 2005]. However, it can be argued that these response systems are also capable of constituting a perceived threat to others. Furthermore, contrary to official Russian, Chinese and U.S. statements, space weapons tests do occur. In early 2008, a U.S. spy satellite was shot down by U.S. forces, officially and very plausibly for safety concerns regarding the toxic hydrazine tank of the spacecraft. The test was announced several days before the incident and was extensively documented with original video footage [Oberg 2008, Garamone 2008]. China, too, destroyed one of its own satellites in early 2007 with a ground-launched missile, creating debris in an orbit reportedly crossing the ISS [Covault 2007]. The Chinese operation was much more secretive and remained officially unconfirmed for days after the event, allowing unnecessary international tension to mount. The sequence of these events could also be interpreted as one country trying to reproduce the previous successes of another country and suggests that any militarization of space will be emulated by other international space powers. The reason, at least for the U.S. satellite destruction, was likely a safety issue, but was significantly diffused by media interpretation. Independent of potential additional reasons, the tests in both countries provided a unique opportunity to actually test ASAT equipment on a real target in space. And if aging spacecraft with similarly toxic contents are deorbiting in the future, this test also proved that such threats can be deflected, even with hardware not specifically developed for the task.

These incidents could be interpreted as simple ad hoc adaptations of existing military technology without separate development efforts. But a satellite hit is not a trivial operation, and requires properly developed and researched hardware as well as careful calculations. This, in turn, requires proper funding and sufficient planning time. The aforementioned U.S. missile launch cost an estimated $100 million, not counting any political fallout [Samson 2008]. It is thus possible that there are other space weapons technology developments underway in various countries that are submerged in classified compartments. Furthermore, the fact that U.S. military spending has expanded as a result of recent international conflicts can have one of the following implications for space weapons funding. On the one hand, there might be insufficient funds left to work on ambitious technological development projects, especially when taking the effects of the current international economy into account. On the other hand, there could also be a significant monetary surplus in various military departments which would allow for the development of space weapons without using directly and publicly attributed resources. The occasional ASAT tests in different countries suggest a not negligible probability for the second option. Against the background of such recent tests, Hitchens *et al.* proposed a novel bargaining strategy to encompass wider policy issues to achieve an improved security for all space assets, aiming at a novel code of good conduct in space [Hitchens 2008].

The space security technologies discussed in this chapter are based on the U.S. proposals outlined in the United States National Security Space Management and Organization plan [Andrews 2001, Weiner 2005, USAF 2003]:

- Military space planes (MSP), including reusable space operation vehicles (SOV) and fast re-entry vehicles dubbed common aero vehicles (CAV);
- Tungsten carbide-uranium cermet rods, which could be installed in the Earth's atmosphere and strike the surface at high velocities;
- Space blimps (Tactical Redirected Energy Initiative and Aerial Relay Mirror System, ARMS) as mirror carriers for ground-based laser reflection and aerial laser relays (the U.S.A. commenced research on space-based lasers, but later mostly discontinued it [Elhefnawy 2006]);
- Radio waves as weapons, such as those investigated in the High-Frequency Active Auroral Research Program,(HAARP);
- Increasing offensive capabilities of satellite technology that could arise from the development of autonomous proximity vehicles (APV), which have been claimed to be capable of disrupting other satellites [David 2003].

While not all these proposals have been carried into actual public project phases, they will be described and assessed in detail below as a means of assessing their technological feasibility.

1. Space Planes

The U.S. (and other countries') military has long been interested in the development of MSPs [GS 2008, Bertin 1996, Pike 1999]. This interest, together with civil programs, resulted in a large variety of proposals for reusable launch vehicles (RLVs). Despite incurring serious costs, none of these programs, with the sole exception of the Space Shuttle, has resulted in a final product [Bertin 1996, Pike 1999, Bradford 2004, Bekey 1994, Baird 1996, Freeman 1996]. One of the program advancements was a SOV, a reusable strike platform assembled from Space Shuttle and fighter aircraft components. SOV payloads could be small surveillance or replacement satellites or other craft such as a CAV. The CAVs re-enter the atmosphere to conduct a strike mission but are not necessarily limited to deployment from an SOV, and can potentially be started from missiles. During re-entry, these craft protect their ordnance until they deploy it on a target.

Response times to a global target of less than 2 h are envisioned for CAVs [Pincus 2005]. The strike platform has global reach and a faster response time than ballistic missiles. A CAV launch is also difficult to detect as it comes from a mobile high-altitude platform and produces a smaller heat signature than ground-launched rockets. Although there is currently no publicly declared funding for on-orbit CAVs, it is technologically possible to place them constantly in space and deliver them at any time without having to use a SOV launch platform. The political consequences of permanently stationing weapons with almost undetectable deployment signatures would be severe, however. In 1998, President Clinton line-item-vetoed development of the CAV and any mention of similar weapon systems and Congress funds for CAV development were frozen for one year [Bonsor 2000]. The veto was later overturned and NASA continued working on reusable non-military space planes. The Bush administration pushed the development of CAVs aiming to produce operational CAVs by 2010, but this aim has since been revoked.

With too many political controversies surrounding stationing CAVs permanently in orbit, the U.S.A. focused on the deployment of CAVs by means of carrier aircraft. This makes sense as the response time is scarcely larger for SOV/CAV compared with orbital CAVs. If larger numbers of the re-entry vehicles are required, this approach can supply them. There are also technical concerns about installation of CAVs in orbit. More propellant is required for the inevitably longer distance into stable orbit and back. Additional space will be taken up if the internal systems are shielded against the space environment. This adds to the mass and leads to the requirement for even more on-board propellant. Re-entry heat is generated as a function of vehicle mass and speed and increases the difficulty of successfully carrying out a mission. Alternatively, space radiation-hardened materials could be used in the systems, but they have to be developed specifically for such applications and would significantly increase the cost of CAV control systems and ammunition. The chemistry of components such as explosives, ignition charges and lubricants would also have to be adapted for long-term space stability. Solutions to all these problems would be possible with significant monetary expense, but seem unnecessary in the face of insignificant response time benefits and political opposition.

It is disputable whether the CAV actually infringes international treaties. Strictly speaking, as there is no treaty forbidding the stationing of conventional weapons in space, CAVs could be created and deployed [USDS 1967]. The SOV/CAV does not necessarily carry weapons of mass destruction (WMDs), does not permanently remain in orbit, must ground for refueling, can be operated remotely and robotically and is not a heavy bomber as defined by the Strategic Arms Reduction Treaty (START). These systems would circumnavigate previous international treaties without violating them. Technologically, however, once the system is developed, there is nothing to stop owners from switching the CAV ordnance to WMDs.

CAV supporters have already proposed plans for the post-development of the first-stage SOV-based CAVs. Once MSPs become standard technology, the second phase of development would see the deployment and permanent installation of CAVs in orbit with attached orbital support and deorbit packages [Phillips 2003]. These systems would be time-limited as fuel and power restrictions create a finite system lifetime, on the order of several months, after which they would probably be deorbited unarmed. Third phase planning includes a shuttle bus containing and sustaining several CAVs. The shuttle would be permanently stationed in orbit transporting systems developed for prolonged stationing in space. The busses could conduct limited maneuvers to confuse surveillance systems without exhausting the power level of the individual CAVs. Technological obstacles to the first CAV phase are mainly related to funding. The U.S. 2001-2009 administrations seemed inclined to press for development in that area at the cost of other national developments, from health care to alternative energies, especially during its first term [Aleklett 2010]. Ultimately, these efforts did not result in the development of a usable product.

Severe international repercussions can be expected from even the initial stages of CAV development. Even if early SOV systems were developed and accepted by the international community, they would inevitably provide the first step for the constant installation of arms in space. The high cost and uncertain reliability of the system remain to be resolved. The difficulties with established Space Shuttle technology (retired in 2011) and the Russian Buran program (discontinued in 1993) over the past few years have demonstrated that reaching LEO is still not a safe routine. Or, as Chairman of the Heritage Foundation Missile Defense Study, H.F. Cooper, put it: "The U.S. has invested about $4 billion over the past four decades…not counting Shuttle development and operations. The residue of our total investment is four (now three) aging Shuttles, one crashed vehicle (now two), a hangar queen, some drop-test articles and static displays" [Cooper 2001].

Cooper's views represent the U.S. Air Force's desire for a reusable spacecraft capability aiming at reducing the cost per kg cargo from $20,000 to below $2000 [Grant 2001]. This would also help to invigorate the commercial spaceflight markets [Cáceres 2004, Harris 2008, Elias 2001, Bugos 2008]. NASA had been involved in R&D of a non-military space plane as a successor to the aging Space Shuttle and took a renewed interest in it after the Columbia accident and the subsequent troubles with Discovery on orbit [Hall 2003, David 2003b, Jonson 2003]. Most of the past projects under investigation were discontinued because of either engineering obstacles or budgetary restrictions. An analysis by Broniatowski *et al.* found that in order to maintain a successful space program, annual funding for space technology must take into account the sustainability of existing projects and not solely focus on new projects [Broniatowski 2008]. With the Space Shuttles retiring in 2011, and follow-up programs such as the Orion spacecraft and Project Constellation cancelled or financially restricted, the only currently available space reentry vehicle is the Russian Soyuz system [Montluc 2010]. Military development of reusable space planes could provide a benefit for civil space travel as well, even though the U.S. Department of Defense (DoD) appears to be hesitant to work with NASA as the aims of the two agencies appear to diverge. Its dependence on NASA has left the DoD with no space plane to date. Development of the technologically superior one-stage-to orbit spacecraft (like the X-33) was pursued by both agencies and Boeing until stress tests on a hydrogen tank led to catastrophic failure of the tank composite shell layers [Grimsley 2001, Gates 2006, Berry 2001, Duffua 2008]. The project was cancelled; it provides a good example of how technological and fiscal issues can successfully impede development [Deptula 2008]. The problems with X-33 are, however, merely materials engineering-related and could be resolved with significantly lower budgets than for starting an entirely new program.

Space Planes– A Summary

The development of military and non-military space planes is financially and technologically feasible. It remains unlikely that such technology will be used to permanently station weapons in space. The possibility of converting the current offensive nuclear threat to an active defense system is possible, but unlikely to be accomplished by these technologies. The technology for the successful development of space planes is available today, and, in light of the retired Space Shuttle technology, overdue in implementation. It is likely that, had all previous spaceplane system design studies been performed within one agency and with a focused development plan, space planes would be an accepted reality today. However, with a strong political and economic will, such modern reentry architecture could still be established in the near future. Private sector enterprises are currently setting up to exploit the business opportunities left behind by the retired Space Shuttles and the discontinued Buran program.

2. Ceramic Rods

Among various proposals for space-based mass-to-target weapons, installing ceramic or cermet rods in an orbital launcher platform actually received some funding and significant media interest. Such rods would be released and sped up to hit any target worldwide in less than 2 h. No space-based strike weapon has yet been tested or deployed in space as there are major engineering and political obstacles to such systems.

Ceramic rods are solid rods made mainly from tungsten carbide (WC), either as a single phase material or as a cermet mixed with other metals such as uranium or cobalt [Du 2005]. These darts are able to withstand atmospheric re-entry at high velocities and steep angles with minimal erosion thanks to their high liquefaction temperature. Because of their high-impact velocities, they do not require any warhead, but convert their high kinetic energy into thermal and deformation energy upon impact. Since kinetic energy is proportional to the square of the velocity, it becomes clear that the ultimate gain is to maximize the rod's terminal velocity. This implies high orbits to have sufficient space to accelerate, but high orbits reduce system response time and accuracy. Furthermore, high velocity projectiles have a tendency to detonate above the surface rather than bury themselves in a well-defined shaft, this penetration being one of the advertised hoped-for properties of the rods. Impact properties depend on the rods' mass, impact angle, thermal energy and velocity as much as on the shape, microstructure and presence of defects in the material used [Llanes 2002]. Peak performance has been calculated for an optimum range of deorbiting acceleration, impact velocity, area coverage, and re-entry angle accuracy, yielding an orbit altitude of 8000 km, and response times to target of up to 2 h [Spacy 1999]. The speed and lack of critical vulnerable parts makes this technology practically invincible once deorbited. This latter property could force other countries to develop space-based counter weapons in order to avert the threat imposed by such high velocity rods.

The system proposed consists of two satellites in close proximity, one serving as the operation system and one as the launch platform. The same technological limitations as for any equipment stationed in Earth orbit apply: the electronics must be radiation-hardened and the equipment must be capable of withstanding the space environment. The rod bundles also need deorbiting acceleration modules, which would have a finite lifetime in space. The rods themselves are rather simple technology. WC is widely used as a hard material in cutting tools, pulp and paper mills, etc., and is not too detrimentally affected by vacuum or by cold temperatures. However, the rods still have to survive re-entry and thus have to be of a specific shape that does not create too much atmospheric friction and allows air to continually cool the rods. Superheated material can evaporate non-symmetrically from the falling rods, changing its aerodynamic properties and altering the drop path, thus making the system imprecise.

In order to survive the heat of re-entry, the dart dimensions have been calculated to be at least 7 m in length and 30-50 cm in diameter. WC has a density of about 16 g/cm^3, 50% higher than that of lead, with the consequence that these darts are heavy. It is thus financially and technologically near-impossible to ship sufficient quantities into space with current launch technology [Hiriarta 2010, Petronia 2010].

From a political point of view, ceramic rods and any other orbital weapons affect all aspects of international space security. Countries with space-based strike systems such as cermet rods could limit other countries from accessing space. This could be from the earliest stages of launch, by simply destroying space launch capabilities [Collard-Wexler 2006]. Similarly, every nation's space assets would be endangered. While the recent intentional destruction of satellites by the U.S.A. and China proved that ground-based technology is capable of taking out space assets, space-based systems could do so with further reach and shorter response times. Since rods could also be used as fast-strike weapons on any global target, they present a significant security concern to every nation. Beyond this, they can be argued to directly contradict stipulations on the peaceful use of outer space like those in the Outer Space Treaty [Hays 2002, Waldrop 2004].

These challenges alone underline the inherent financial, technological, and political disadvantages of developing space-based strike weapons. But should a decision be taken to develop such weapons, R&D of eroding rods would be unlikely to result in any usable product. Furthermore, such systems could present a rather interesting opportunity for adversaries. Such fractions could use similar systems against the country that deployed the system, especially those that do not have large Earth-based military equivalents [Hardesty 2005]. As outlined earlier, hypersonic projectiles are difficult to defend against. Therefore they will have a much more significantly detrimental effect if used against a country (such as the U.S.A. or other nation with similar military and technology potential) which has allocated significant resources to conventional defense technology in its homeland, all of which would be rendered useless [Preston 2003]. This would allow possible adversaries a significantly larger increase in combat power than it would for the U.S.A. or countries with similar development plans. No installation of weaponry in high orbits with sophisticated control equipment needs to be developed and deployed by groups intending to harm such countries [WH 2003]. Conventional weapons can be placed in LEO, without the need for steep re-entry angles and high-impact velocities, and could be placed directly above any given country without the need for global coverage. This reduces response times to below half an hour. Transfer and installation of conventional weapon systems can be accomplished with proven return capsule technology and would thus be available to a large number of countries.

The scenarios outlined above are made possible if any country starts developing and deploying offensive arms in space. Either the technology is replicated by other countries that can subsequently bypass the first country's defenses, or adversaries simply answer with a comparably low-tech approach that can saturate or overwhelm defense systems. Both possibilities leave countries such as the U.S.A. more vulnerable than before these developments. Furthermore, all investments in missile defense systems to date would be doubly futile, as current systems do not reliably work against conventional ballistic missiles, are controversial against short range missiles [Postol 1992, Zraket 1992], and are even less effective against space attacks for example with cermet rods.

Some analysts argue that if the U.S.A. takes control of space first, it can deny access to all other countries and these scenarios can be avoided through the dominance achieved [Smith 1999]. The simple vastness of three-dimensional space around the globe, however, makes it very challenging to control access to space from all sources and attempting to enforce such control would be highly problematic, leaving aside the political implications of such actions. The same would be true for any nation with similar aspirations.

As an alternative to space installation, missile-based ceramic rods have been proposed. These systems would not face the technological and political implications of space-based technology. These would be tactical missiles with a metal carbide cermet head and no warhead, only slightly less effective than space-based rods because of reduced terminal velocity, but with almost the same response time and significantly lower development costs. These weapons might not contribute to what some countries perceive as an inevitable arms race in what is described by Honorable Duane Andrews *et al.* as the "new battlefield space" [Andrews 2001].

Ceramic Rods – A Summary

In addition to the international security concerns outlined, the technical feasibility of space-based cermet darts remains questionable. There is very little public information on actual research of these systems. The official removal of funding for related projects might in this case be a result of the realization of the futility of researching these systems. While such research could have been submerged into unofficial channels, the fact that the resulting systems are unlikely ever to work properly makes this assumption doubtful. Today's launch structure makes it very expensive to lift even small quantities of heavy and bulky objects into space. The systems have an intrinsically low precision because of the inert drop pathway, difference in atmospheric densities and turbulence and uneven surface liquefaction on descent. But even if these heavy systems could be lifted into orbit in sufficient quantities to obtain global coverage and the desired high precision achieved, the projectile impact properties are still uncertain. Although wishful thinking has them as deep impact devices, hypersonic superheated projectiles might simply detonate above ground [Preston 2003]. This can already be done today and at significantly lower cost with proven missile technology.

3. Space Lasers and Laser Blimps

Space and airborne energy weapons have long since been popular within some circles. The prospect is interesting: destruction of any selection of targets in orbit or on Earth, including mobile targets such as satellites or incoming missiles, with a simple short burst of energetic electromagnetic or particulate radiation. Such a ray can be created by photon/light amplification by stimulated emission of radiation (PHASER/LASER), by microwaves or particle beams [Varni 1996, Scully 1992]. The technological downsides are also well known [Spacy 1999]. In order to create a signal significant enough to disrupt solid matter, the power generators and other internal systems have to be much larger and heavier than today's space transport infrastructure allows. Pinpoint focusing remains difficult, as even the best ray generators will output a slightly divergent beam, a problem made more severe by the huge distances the ray has to cross for a global strike. Furthermore, atmospheric density fluctuations can distort and diffract radiation. However, this is a matter of physical calculations and can be accounted for [Hardy 1994]. The internal cooling systems create slight vibrations, which, multiplied by the distance, make it difficult for a beam to properly focus on a designated target, especially if it is mobile. This restriction already limits solidly mounted ground-based lasers; it can be assumed that satellites will create vibration that will make them imprecise even at a short distance of only several dozen to several hundred kilometers. The cost for the development of these satellites was estimated in 1998 to about $81 billion, but this investment does not automatically imply that the resulting technology will function satisfactorily [Possel 1998]. Also, cost per target destruction is around $100 million. This is several orders of magnitude higher than conventional strike methods [Borger 2005, DeBlois 2005].

In order to reduce cost and the total number of systems necessary to cover the entire globe, analysts have proposed replacing some space lasers with bifocal mirrors. Bifocal mirrors are two orthogonally arranged telescopes, catching and refocusing an incoming ray and then redirecting it through the second telescope. This reduces overall cost and ballast to orbit at identical area coverage. These systems have been assessed to be cost-effective and deployable within the next few years [Possel 1998]. Nevertheless, the same source also mentions that this technology has never been tested before and that the laser emission systems still have to be developed and shipped to space.

Another option proposed are ground-based ray emitters whose beams would be redirected by mirrors that are either installed in orbit or carried around by moving aircraft such as zeppelins [Wilson 2002, DOD 2003]. Being ground-based implies that energy is accessible in sufficient abundance and total mass and system dimensions do not matter significantly. But it also means that the rays have to traverse longer distances in atmosphere where airborne particles or precipitated liquids can scatter the emitted radiation. Furthermore, projects such as the Airborne Laser have already seen lasers mounted directly on a variety of aircraft, including zeppelins, none of which has yet become operationally useful. Consequently, all related long-term projects have been scaled back or cancelled [Schachtman 2004]. The focus of R&D activities shifted to relaying beams via mirrors rather than directly mounting energy beam weapons on aircraft. Such 'mirrored' aircraft were developed by Boeing under government contract and by Lockheed Martin in an independent R&D airship testbed study that accomplished its first flight test in February 2006 [GS 2006, Dornheim 2006]. However, most of the R&D craft are far from finalization and the suspension of relay mirrors from airships will introduce significant vibration and imprecision, unless costly countermeasures are engineered into the relays.

Required power output ranges from 25 to 130 MW for space-based geosynchronous relay mirrors and could be reduced for blimp mirror carriers [Varni 1996, Goldwasser 2006]. In other words, for every eight earth-based laser emitters constructed, an average-sized 1 GW (e.g. nuclear) power station would have to be built to provide the necessary power [Longwell 1995, Evans 2007]. Admittedly the energy is usually only required for micro-bursts, not continuously, but the availability of such power in space-based systems is more than questionable. Research breakthroughs in laser focusing, beam jittering and mirror reflectivity are still necessary to make this technology successful. The large distances involved, up to tens of thousands of kilometers, create difficulties that are unlikely to be overcome because of the physical limits of the systems. To be effective, beam divergence should be below 0.3 s of radiant (or 0.08 mrad) on a target [Varni 1996, Goldwasser 2006]. Such a laser has not yet been developed; current cutting-edge research trials aim at 1 mrad precision [GS 2006]. Particle or ion emitters usually have tighter emission divergences, but charged particles are more easily distorted in the atmosphere and deflected by the Earth's magnetic field, gravity and atmospheric charges.

On the other hand, recent developments have led to lasers with unprecedented 15% efficiency. In this case, 'only' 85% of the power input is converted to heat, compared with 99% heat losses in standard chemical lasers. Novel efficient diodes and microelectronics, increased remote power production and storage capabilities and the introduction of low-heat-loss materials such as ytterbium and >99.99% reflective surfaces improve on these issues [for example Kammen 2006, Goldwasser 2006, Gloss 2008, Fulghum 2005, Wu 2007, Aurbach 2005, Raffaelle 2005, Malenfant 2006, Rosati 2009, Rose 2006, Walter 2002]. Reflection is nothing else but excitation and de-excitation of the electrons present in any material. If incoming radiation is not reflected, the energy is absorbed by the material as heat and can harm the carrier material itself. Thus it could be assumed that increasing the reflective properties of targets will protect against energetic beams. But increased reflectivity of the target surface can only partially protect against damage, since material reflectivity decreases with an increase in radiation wavelength (for example direct heat, or infrared radiation). Likewise pulsed laser technology can potentially remove reflective surfaces by the transfer of direct momentum to the molecular surface layers leading to spallation. However, this could also destroy reflective surfaces on mirrors directing the beam at a target. Alternatively, an intense radiation burst can be fired into the direct vicinity above target missiles in order to create a shockwave that could potentially stop a missile. While this has been shown to be effective on small craft and for negligible distances, it is untested on any larger range.

Focusing the beam also requires large optical lenses and high reflective mirrors, which creates another problem related to materials engineering. Every material possesses intrinsic defects and, as the total size increases, so does the occurrence of these defects, adding to the imprecision of focusing [Walter 2002, Harr 1989]. Development and production of high-quality low-defect surfaces with a large area will certainly delay the creation of lasers as global security technologies.

Space Lasers and Laser Blimps – A Summary

The idea of having space-based energy weapons is interesting as their time to target is short. Unfortunately, energy efficiency and precision are low. Energy beam divergence reduces the energy per unit area that arrives on target to a degree that makes them potentially insufficiently powerful over the distances involved. Beam divergence can be and has been reduced by research, but there is a physical limit to this reduction. Divergence can never be totally removed. Size and mass of the system is further increased by the cooling system for the high-energy beam emitter, possibly to a point at which the systems have to be assembled on orbit. All on-board systems from power and beam generation to cooling introduce certain vibrational frequencies that can cause substantial oscillations on target. This is especially true for space systems, which are drifting without any solid support that could reduce the effect of vibration on beam deterioration.

While ground systems are still subject to the same physical limits described above, they are less subject to vibration, power is more readily available and the heavy, complex system does not have to be lifted into space. On the other hand, the beam traverses longer distances in a distorting atmosphere. This can mostly be addressed by using mathematical models for pre-distortion of the beam. Blimps in general offer a great carrier capability and the usage of suspended mirrors would alleviate the need for space-based beam relay stations. The sustainability of the reflective properties of relay mirrors during radiation and the effectiveness of any reflected ray on-target would still have to be demonstrated.

Energy beam research for the military has produced a large variety of interesting technological advances for the scientific community, but seems unusable for high-precision long distance applications in the foreseeable future. Spreading of the beam while it traverses several thousand kilometers and the consequent loss of specific energy is probably the most severe physical limitation of energy beam weapons, rendering the systems practically useless over long distances. For these reasons, while space lasers were at some point an actual research program, most of the research has been largely discontinued, but continues in few small scale projects.

4. Radio Waves

Experimentation with the ionosphere, the charged protective layer that surrounds the globe and deflects interstellar and solar high-energy radiation, is another topic that has a long history in space research. Research started with mainly U.S. and Russian nuclear fission and later fusion explosions that permanently altered Earth's ionospheric layers. In 1963, the U.S.A. injected thin 1.78 cm long copper needles into an Earth orbit, attempting to create an ionosphere with the needle dipoles that would allow for global radio transmissions [Lovell 1962]. After initial experiments, the project was discontinued as communication satellite relays improved and international pressure over the significant creation of space debris, which lasts into the present day [Wiedermann 2001, Barhorst 2000], was brought to bear. Later experiments with Skylab established a sense of how the plasma layers in this part of the atmosphere can be harnessed for signal transmission and interruption [Mendillo 1975, Bertell 2006].

Radio frequency (RF) waves have been researched for a long time. Possible applications are underground imaging, chemical elements tracing in the ground, or underground shelter detection [GS 2006]. With certain modulations, a high-energy beam directed into the ionosphere by a ground-based antenna can create an Artificial Ionospheric Mirror (AIM) and alter the plasma layer structure in the atmosphere [Koert 1990]. These layers can be used to deflect RF energy emitted from the planet's surface onto another part of the globe. When the layer generation is altered in phase and frequency, the layers can (up to a point) be tilted and refocused at continually higher altitudes.

These activities are some of the focuses of diverse research carried out at HAARP. Proposals range from generating extremely low-frequency waves for communication with submerged submarines to the understanding and exploitation of ionospheric properties. Application of beyond-the-horizon radar systems with higher precision than current devices and the capability to detect anything from underground structures (area tomography) to incoming aircraft are further hoped-for outcomes of the HAARP research. A potential outcome of the research could be the alteration of Earth's geomagnetic field to deflect and disrupt foreign communications and to transmit and amplify hard electromagnetic radiation to a large area anywhere on Earth. This would offer advantages to any military applying such technologies, if they could be realized [GS 2007]. Intentionally wiping out communication over a large target area by RF radiation from the ground is a novel proposal to replicate the effects of atomic explosions in the atmosphere. Military systems could be hardened against the electromagnetic effects by proper materials engineering and could continue to function. Alternatively, radiation could be used to induce electromagnetic pulses to disrupt or permanently damage satellites [Begich 1995]. However, even assuming this research is still ongoing, military applications have not been proven and the magnitude of the impact is likely to be small compared with the vastness of the ionospheric layers.

Besides the obvious military advantage of determining the strength of enemy underground shelters, the underground wide-area tomography could also have a positive impact on peaceful non-proliferation surveys. It could detect structures that are invisible to today's satellites [Zimmerman 1990 and 1991]. In addition, it could be used to find and estimate natural resources on a grand scale, thereby significantly reducing the cost of such missions. However, there are scientific worries about the wavelengths used in such experiments as they might potentially be similar to those used to activate heat transducing proteins on a sub-traumatic level [Begich 1995, ONR 2004]. While these concerns might be valid, this effect has not yet been scientifically confirmed, and seems unlikely to occur over large areas and long distances.

There might, however, be detrimental effects to Earth's atmospheric layers resulting from this research that could last beyond this century. The early high-altitude fusion experiments have already changed the Earth's protective layer, possibly for millennia, by changing the patterns of the electron layers in the ionosphere [Begich 2005]. The effects of further tampering with the atmospheric layers that protect us from solar and stellar radiation remain largely unknown [Steigies 2000 and 2002, Baker 1985, Sojka 2002]. This must be analyzed in context with the scale of the layer in question, and the total amount of energy that can be directed at it from Earth-based systems. First, most of the experiments have to be conducted during the night [Thiemann 2001]. During the day, the Sun's radiation negates most human-induced changes. This electromagnetic radiation irradiates these layers with a much higher intensity and consequently levels out most human-induced changes. Thus effects lasting more than a few days are highly unlikely, barring energy inducements on the scale of fusion explosions. There is no consensus on the severity and longevity of these effects, but generally care should be taken as effects in our atmosphere will have a global impact that might damage not only a potential adversary but also the emitter of the radiation.

Radio Waves – A Summary

The prospect of harnessing atmospheric layers for communication as well as for disruption of foreign communication is interesting. Very expensive submarine communication as well as currently ineffective ballistic missile defense systems would be rendered obsolete, to be replaced by a simple, effective system, shifting the world's political power balance towards a defensive mechanism. This also opens the possibility of substantial financial savings if this technology can be harnessed successfully to these ends, which has not been shown to date. The money and time required to actually achieve these goals are unknown. The power necessary to alter ionospheric properties on a global scale precludes space-based assets. Whether sufficient power can be supplied in one ground-based system is unknown. Detrimental long-term effects of tampering with the ionosphere that protects us from radiation cannot be ruled out, but have never been proven. High-altitude nuclear tests of the 1960s introduced new electron layers and permanently altered the flow patterns of the ionosphere. However, the ionosphere is under heavy radiation while exposed to the Sun. Human influences are comparably tiny in scale. The intentional unselective disruption of all (foreign and national) communication that relies on the ionosphere would have significant negative effects on society, but is not likely to be technologically feasible with ground-based antennae. It is unlikely that significant military strike options will result from any of this research.

5. Satellites

The use of satellites for military purposes is far from new. Since the early 1970s, the U.S.A. and Russia have deployed geostationary early-warning satellites at a height of >35,000 km, in the U.S.A. under the Defense Support Program. This rather old technology was planned to be phased out by 2010, and successor technology is in development. Six Space-based Infrared System (SBIRS) satellites, with two in high elliptical orbits and four in geostationary orbits of different altitude, are planned to be deployed as upgraded replacements [Jasani 2001]. SBIRS testbeds seem to perform well [Jasani 2001]. However, the budget of $12 billion, a total overrun of $2 billion has troubled the deployment and pushed the SBIRS program completion date beyond 2014 [Tillinghast 2005, Morris 2005]. Satellites with either infrared (IR) or radar sensor capability are comparably old and proven technology, and have been employed by both Russia and the U.S.A. for decades and are probably in use by most other space capable nations, too. The current U.S. systems technologically surpass the Russian craft as Russia has faced financial problems with its space technology, creating a perceived shift in equality of arms between these two nations.

A different approach to space security technology is followed by the deployment of autonomous proximity vehicles (APVs), usually microsatellites with a mass of less than 150 kg [Lewis 2004, Boeing 2006]. These tiny satellites have several advantages over their currently rather bulky peers in orbit. First, transport into space becomes easier and cheaper, and dozens of these microsatellites can be ferried to space together and employed faster in larger numbers. Maneuverability in space is also much in favor of smaller, lighter objects. In order to carry out maneuvers, spacecraft need thrusters that require fuel. The need for propellants increases with an increase in the intrinsic mass of the satellites and the acceleration required for a set of maneuvers. The need for propellant can be expressed as the ratio between the satellite mass and the propellant mass required to accelerate the craft to a certain velocity, not taking into account that the propellant itself increases the mass of the satellite. The following approximation has been calculated by Wright *et al.* [Wright 2005] using the example of a craft with an inert speed of 3 km/s. For maneuvers requiring a change in velocity of 2 km/s, the craft would have to carry 0.9 times its own mass in propellants. Twice these maneuvers (total change of speed 4 km/s) would require 2.8 times the craft's mass in fuel. And a velocity change of 8 km/s would require the transport of 13.4 times as much propellant as craft mass. Such large velocity changes or in other words so many maneuvers are very unlikely in practice, however. Nonetheless, the same limitations apply to all spacecraft, including missile systems. This example simply emphasizes the need for development of smaller, lighter craft, with the additional potential for significant cost reduction.

Air Force officials claim that developed microsatellites are mainly meant to be used for satellite servicing, celestial body sample return delivery, and cargo delivery to space stations. While there is no doubt that that is true, APVs could also be capable of disrupting the functionality of other satellites without necessarily destroying them entirely, but this can be true for any technology moved into orbit. APV usage could be even more versatile, ranging from defending spacecraft to actively jamming or intercepting foreign communication signals of any source, disrupting functionality of other satellites or just passive close-up scanning of satellites with a resolution unsurpassed by ground-based sensors and scanners [Kramer 2001, Partch 2003]. It should be added here that a similar case of military and civil 'dual use' can be made for all space technology, not just APVs. In a 2006 publication by the UK Parliamentary Office of Science and Technology, the authors mention that space is used extensively for military purposes but that there are no weapons in space and elaborate that many technologies developed for peaceful or defensive purposes could also have offensive uses [Nath 2006]. Current satellite technology allows these satellites to accomplish remote operations with supervised autonomy (ROSA) and to navigate in the direct close proximity of other satellites [Singer 2005].

Tests with a NASA APC dubbed Demonstration for Autonomous Rendezvous Technology (DART) were not entirely successful as the microsatellite accidentally rammed a test object in space instead of circumnavigating it [Bender 2005]. This DART mission lasted a total of 24 h and cost about $110 million [Young 2005]. NASA seemed at this point to no longer consider robotic satellite servicing a valid option and has reduced its involvement since [Bender 2005]. Close proximity technology is still immature, but one of the last versions of a U.S. Air Force APC, the Experimental Small Satellite (XSS-11), operated for several months and showed great promise and significant technological advances compared with its own predecessor, the XSS-10, which was only designed to function for 1 day [Oberg 2005, Hitchens 2008b, Toso 2004, Page 2006]. Demonstration costs for XSS-11 were estimated at $80 million. The satellite operated in the proximity of 10 space vehicles and no glitches occurred according to published accounts.

The idea of autonomous space vehicles is intriguing. Being able to transport and deliver many microsatellites hooked up to only one delivery ship would significantly reduce the cost and time of satellite space deployment [Lloyd 1999]. International projects such as the German/Russian TECSAS chaser spacecraft (discontinued September 2006) and follow-up missions such as the Deutsche Orbitale Servicing (DEOS) show that remote servicing is a modern and necessary novelty for space technology development that is pursued globally [Ning 2007, Gronland 2007]. Another project advancing this technology is funded by the U.S. Defense Advanced Research Projects Agency (DARPA). This orbital express demonstration program was led by a consortium of companies including Boeing, Ball Aerospace and the Canadian MD Robotics. They developed a >700 kg serviceable commodities satellite (NEXTSat/CSC). This satellite was a testbed for on-orbit refueling, system upgrading, damage assessment and repairs. The project's Phase II flight tests were completed and NEXTSat performed an on-orbit servicing demonstration on an unmanned ASTRO service vehicle, continued in the Space Test Program. According to Parry *et al.*, this test showed that "...the DARPA contract is a crowning achievement of many years of R&D in an autonomous operation of robotic systems for space" [Parry 2004].

Satellites – A Summary

Precise satellite maneuverability in space has been demonstrated in previous space missions such as Deep Impact, from which the software, command and data-handling technology were adapted to the new challenges of spacecraft servicing [Ball 2004]. However, precision has to be improved for close proximity operations as well as the autonomous decision-making process if these satellites are to be successfully deployed in future, be it for friendly repair or hostile sabotage missions.

For peaceful use, microsatellites and servicing satellites make sense as they decrease the cost of space travel significantly. They also have the potential to enhance the lifetime and prevent total project discontinuation of already on-orbit craft such as the Hubble telescope. It remains to be seen how much of this technology will be utilized for military operations, especially after the withdrawal of NASA from close proximity remote satellite servicing and the reassignment of the ground crews to other projects.

With regard to large surveillance satellites, the current U.S. assets in space are technologically outdated (though not yet totally obsolete) and are scheduled to be replaced after 2014, albeit at substantial cost. The necessary technology is available and just has to be improved and adapted for space.

All current satellites face the drawback of literally zero redundancy. Most current military space systems are too heavy and expensive to allow for multiplicity. The development of microsatellites offers the advantage of creating a network of lightweight, easily deployable satellites that share functions within the system [Clare 2005, Milas 2008]. This also makes them less susceptible to attacks and common malfunctions, as the dysfunction of parts of the network does not interrupt overall functionality, as would the destruction of one sole military satellite. But, far more importantly, the necessity for defensive space weapons would be rendered superfluous.

6. CONCLUSIONS

This chapter discussed space security technology and the possible implications of developing and deploying some specific systems, based on published research and development proposals. There is no single nation driving such development. Several countries have their entire space program based in the military. Russia, with the only publicly confirmed (but seldom mentioned) actual space weapon, an aircraft cannon mounted on Salyut 3, has already shown that it is willing to develop and carry weapons into space [Hammond 2002]. This fact is underlined by the creation of space technology such as the failed Russian Polyus ASAT/miner testbed [Godwin 2001, Kornilov 1992]. It is likely that other countries could use such precedents to point at in the case of their own security technology developments. Even if many of the proposals in the U.S.A. never went beyond the planning stage [ADMIN 1986, Heimack 1986], the mere installation of abovementioned relatively primitive gun turrets in space serves further to underline that merely proposing the development of space weapons in the U.S.A. can easily spark real development elsewhere, leading to an arms race without real cause, especially in the light of the media exposure they typically receive. This once more underlines the need for a new code of good conduct in space [Hitchens 2008, Huntley 2010].

The development of reusable space planes could benefit the public space market as much as a military seeking global strike options. The engineering obstacles can be solved with today's technology but this requires a determined and focused effort that should exclude withdrawal once commitment is made. Building a space plane on previously achieved improvements during projects like the X-33 is estimated to cost several hundred million US dollars. This seems low when compared with an estimated more than $12 billion necessary to develop a completely new spacecraft. The necessary separate development of a CAV is likely to be similarly expensive, but appears to be achievable from a technological point of view.

Energy beam technologies, on the other hand, are not likely to function in the near future, especially over global distances, and funding requirements for these systems would be huge. There are physical limits inherent in energy and particle beam systems that suggest they will remain ineffective as anything other than short-range tools. Although recent advances in blimp technology have shown merit, deflection of laser beams by means of suspended mirrors is unproven and is not likely to yield a functioning global strike system based on the observations presented within the foreseeable future.

Likewise, the installation of heavy cermet darts such as WC/U in the atmosphere appears unfeasible. Such systems are unlikely to yield the desired capabilities they are hoped to possess. Even if the difficulties surrounding orbital transport of heavy equipment and long-term space installation can be addressed, these weapons seem unlikely to perform better than today's ground-based weapon systems. It is also not clear whether these weapons would be physically capable of deeply impacting hardened, buried targets because of their very nature. Hypersonic dense impact projectiles with high thermal energy might detonate upon or even before impact instead of achieving the desired deep ground penetration.

The issues surrounding intentional ionospheric alterations are diverse. Several space experiments have shown that these atmospheric layers can be modulated by human technology. If facilities like HAARP can prove they can do the same on a more sophisticated level from ground-based stations, the implications are far ranging. Most ground-based radar systems, early missile-warning systems and expensive submarine communication facilities, to name a few technologies, would become obsolete and could be replaced by one facility such as HAARP. The possibility of performing tomography of almost the whole globe to a certain depth is likewise interesting for both military and civil applications. That advances in this field could lead to directing electromagnetic radiation to intentionally harm other nations is a possibility that cannot be discounted, although with present technology it is unlikely. None of these speculative effects have been discussed in the scientific peer-reviewed literature. It remains unknown what level of funding and commitment would be necessary to achieve any of these goals. The scale of these ground-based systems is unlikely to have significant long-term impact on the atmosphere and ionosphere since the effects are dwarfed both by daily solar irradiation and by the after-effects of high-altitude nuclear explosions, but a cautious and conscientious research approach is advisable.

Finally, with regard to satellite technology, sensory IR satellites are already, although with a slightly different level of sophistication, employed by Russia and the U.S.A. The U.S.A. is in an ongoing process to renew an ageing set of early warning and surveillance satellites within the second decade of the 21st century. Furthermore, microsatellites are being developed to possess autonomous maneuvering capabilities and have the potential to operate near other spacecraft. This could allow them to determine and even repair damage without the need for expensive human missions or the decommissioning of defective craft. It could also be used to intentionally reduce the functionality of other satellites. The technology is too immature to be useful for either operation yet and NASA, at least, has for now withdrawn its efforts from autonomous robotic servicing projects.

The idea of micro or smaller satellites has merits beyond satellite servicing and espionage, one of which presents an alternative to what some analysts perceive as 'inevitable' space war technology. Today's satellites are huge and costly to produce and orbit. Consequently, there is almost no redundancy and thus a great vulnerability to attack. Some exceptions such as the GPS system exist. They consist of a very small network of satellites. Most other satellites operate as stand-alones and are thus prone to accidents and attacks. A swarm of tiny satellites that could split the functions and tasks of today's huge craft would reduce the risk of loss of satellite functioning [Clare 2005, Milas 2008]. Dysfunction of parts of the network could be easily coped with in a larger network. This would also spell an end to the perceived 'necessity' of defensive space weapons.

These deliberations do not take into account the political implications of an arms race in space. Although this race might not be prevented even if the large global players refrain from setting further precedents, global space militarization will start if spacefaring countries follow through with plans such as those deliberated in this chapter. The stationing of weapons of any sort in orbit will doubtless lead to counter-reactions from other states worried about their own space assets as well their ground security. The U.S.A., which for now holds the majority of space-based systems and which has a heavy reliance upon satellite technology, has most to lose from such an event. Space weapons can be allegorized as technically sophisticated landmines. It has been proven time and again that the deployment of mines does not create security, while the damage to a country's soil, or in this case global space (especially U.S. space), will be present for decades. For these reasons opinions about space weaponry, even within the U.S. military, range from realist pragmatism to idealist notions of 100% security as outlined by Hays *et al.* [Hays 2002].

Realistically, high-technology attacks on current space assets of any nation are unlikely. Groups with an interest in disturbing or destroying satellites are not likely to develop and rely on similar high technology. It is easier to deploy low-tech weapons as simple as ramming objects or explosives. Non-linear threat groups have no interest in keeping other issues in mind such as the consequent increase in space debris resulting from such attacks. It is also unclear where a real threat to space assets might come from today. The most urgent “danger” to date comes from the ever increasing economic pressure on companies involved in space technology. While U.S. companies monopolized the space economy sector for decades, companies from other countries are increasingly participating in the market, reducing U.S. profit margins [Lewis 2005, Choi 2006, Harvey 2004 and 2007, Ware 1988, Fabre 2002, Schaffer 2008]. A good example is the perceived frustration with ESA's and the EU's joint development of a Global Navigation Satellite Systems (GNSS) system, Galileo [Lebeau 2008, Secara 2010]. In this case, the U.S.A. could lose both customers and control over data distribution and in the past it has threatened the destruction of Galileo if countries like China have continuous access to its data during a crisis [SD 2004]. Others have suggested that, rather than using threats, the U.S.A. could gain much from more transparency and a closer partnership with China (and other nations) in its quickly expanding civil space activities [Hitchens 2008, Harvey 2004 and 2007, Johnson-Freese 2006, Logsdon 2008a, Novotsi 2008, Zhou 2008, Logsdon 1998, Montluc 2010, Secara 2010, Hiriarta 2010, Petronia 2010]. China, for example, has managed to emulate the developmental steps of space exploration that took other nations decades within a few years, as demonstrated in the first decade of the 21st century by the first Chinese space-walk.

The political implications of the development and installation of weaponry in outer space, where no demarcation is present, are far ranging and beyond the scope of this chapter [Harris 2006]. If any missile defense system, be it “conventional” or space-based, proves effective against small-scale attacks, several possibilities arise, especially if such a system is developed internationally [Forden 2006, Logsdon 2008b]. International security is shifted from an offensive balance of power by threat to a passive defensive balance of power by effective defense systems, the latter being desirable as they do not rely on imminent threat as peacekeeper alone. However, in case of a non-internationally developed program, this could constitute a tilted shift as there would be no passive balance of power. Countries wielding the defensive systems would be the sole entities protected by such technology. This would result in these countries appearing even more threatening to the remaining international community, which underlines the necessity for global players to jointly develop effective defensive systems, preferably ground-based. Otherwise, a singular approach to space weaponization could prove detrimental to general international security and in particular to the traditionally tedious bilateral and multilateral arms control processes. Over the past years Russia appeared to be less interested in upholding its international non-proliferation treaties as defined in agreements such as START and the Strategic Arms Limitation Treaty (SALT) [USDS 1967, Jones 2006]. For its part, the U.S.A. withdrew from the milestone of international arms control, the ABM treaty, in June 2002 [Hildreth 2005], after declaring it a relic from the Cold War, signed with a nation that no longer exists (U.S.S.R.) [Cooper 1996]. Since the U.S.A.'s withdrawal, Russia has had the possibility of either re-expanding its own significant deterrent or cooperating with the U.S.A. on the development of space weapons disregarding other countries, neither option helping global security. Countries such as China perceive current developments as a direct threat potentially leading them to expand on their nuclear arsenal [Liao 2005, Qingguo 2005, Randow 2005, Holdstock 2000]. The current course of the governments of all major global powers seems likely to lead to a global increase in conflict [Huntley 2005, Hitchens 2002], especially as the general population appears largely indifferent to these developments [Grossman 1999, O'Shaughnessy 2004].

Since ASAT and space weapons are not covered specifically by international treaties [Day 2005], it is time for the development of a multilateral treaty on ASAT and space weapons to create international rules for space-based weapons systems and to significantly reduce the risk of future disasters in space. Only significant individual and collective restraint through international policy, including both formal and informal multilateral negotiations, will allow us to continue to peacefully utilize space, as has largely been possible over the past 50 years [Moltz 2007, Hyten 2003]. Preparatory drafts such as the June 2002 and February 2008 proposals by Russia and China (among other nations) to prevent space weapons have been presented. However, they appear to require further preparation work and more flexibility from all international powers to be useful [Logsdon 1998, Skotnikov 2002, UN 2008, Casini 2006]. This is especially true given that START expired in 2009. Yet, to quote Dr. R.L. Garwin of the Science and Technology Council on Foreign Affairs, New York: "Even if weaponization of space is ultimately inevitable, like our own deaths, why should we rush to embrace it?" [Garwin 2000].

ACKNOWLEDGMENTS

The author gratefully acknowledges the Research Assistantship Award by the Simons Centre at the University of British Columbia (UBC) for this work. The information presented in this chapter does not necessarily reflect the views of the author's host institutes. A modified version of parts of this text has previously been published in Space Policy 24 (2008) 208–223, and is used here under Copyright License 2341780823157 with kind permission from Elsevier.

REFERENCES

ADMIN, (1986). US Administration, Pres. Reagan. *National Security Study Directive NSSD4-86 anti-satellite* (ASAT) options, 20 October.

Aleklett, K. (2010). Peak Oil, Risks and Sweden's plan for mitigation. *Australian Association for the study of peak oil and gas*.

Andrews, D. P. & Horner, C. A., *et al.* (2001). *Report to the Commission to Assess United States National Security Space Management and Organization*, 11 January.

Arcus, P. (2003). American national missile defense system, *Space Policy*, *19*, 7-13.

Aurbach, D. (2005). A review on new solutions, new measurements procedures and new materials for rechargeable Li batteries, *Journal of Power Sources*, *146*, 71-78.

Baird, H. D., Acenbrak, S. D., Harding, W. J., Hellstern, M. J. & Juselis, B. M. (1996). *Spacelift 2025-the supporting pillar for space superiority*. Air Force 2025, *vol. 2,* August [Chapter 5].

Baker, K. D., LaBelle, J., Pfaff, R. F., Howlett, L. C., Rao, N. B. & Ulwick, J. C. (1985). Absolute electron density measurements in the equatorial ionosphere, *Journal of Atmospheric and Terrestrial Physics*, *47*, 781-789.

Ball, (2004). Ball Aerospace and Technologies Corp. *Orbital express demonstration*, www.ballaerospace.com June.

Barhorst, L. J. C. (2000). RAE table of earth satellites (revised edition), *Royal Aerospace Establishment/Defense Research Agency*, Comment: 34.

BBC, (2001). *Russia condemns US missile test*, BBC News, 15 July 2001, accessed 12 December 2005.

Begich, N. & Manning, J. (1995). Angels don't play this HAARP. *Advances in Tesla technology*, 1st edition, Earthpulse Press.

Begich, N. & Manning, J. (2005). HAARP project: disaster or 'pure' research? *Earthpulse Flashpoints*, *3(1)*, 1-11.

Bekey, I., Powell, R. & Austin, R. (1994). NASA studies access to space, *Aerospace America*, *32*, 38-43.

Beljac, M. (2008). *Arms race in space? Institute for Policy Studies*, Washington, DC, Foreign policy in focus.

Bender, F. (2005). *NASA's DART spacecraft bumped into target satellite*. Space News www.space.com, 22 April.

Berry, S. A., Horvath, T. J., Hollis, B. R., Thompson, R. A. & Hamilton, H. H. (2001). X-33 hypersonic boundary-layer transition, *Journal of Spacecraft and Rockets*, *38(5)*, 646-657.

Bertell, R. (2006). Background of the HAARP Project, *Earthpulse*, 17, February.

Bertin, J. J., Boehm, J. M., Matthews, S. B., McIntyre, T. C., Rasmussen, B. L. & Sitler, A. R. (1996). *A hypersonic attack platform—the s3 concept*. Air Force 2025, vol. 3, August, Chapter 12.

Billings, L. (1997). Frontier days in space: are they over? *Space Policy*, *13(3)*, 187-190.

Boeing, (2006). *XSS micro-satellite information* www.boeing.com/defense-space/space/xss, accessed 20 January 2006.

Bonsor, K. (2000). How space wars will work—space weapons in development www.science.howstuffworks.com/space-war2.htm, June 2000, accessed 13 February 2006.

Borger, J. (2005). Bush likely to back weapons in space, *The Guardian*, 19 May.

Bradford, J. E., Charania, A., Wallace, J. & Eklund, D. R. (2004). *Quicksat a two-stage to orbit reusable launch vehicle utilizing air-breathing propulsion for responsive space access*. In: AIAA Space 2004 conference, paper 2004-5950. San Diego, CA, September.

Broniatowski, D. A. & Weigel, A. L. (2008). The political sustainability of space exploration, *Space Policy*, *24*, 148-157.

Bugos, G. E. & Boyd, J. W. (2008). Accelerating entrepreneurial space: the case for an NACA-style organization, *Space Policy*, *24*, 140-147.

Cáceres, M. (2004). For space markets, greater volume remains elusive, *Aerospace America*, 13-14, September.

Casini, S. (2006). Dealing with the international implications of space exploration, *Space Policy*, *22*, 155-157.

Cebrowski, K. (2004). *Statement before the Subcommittee on Strategic Forces Armed Services Committee*, United States Senate, 25 March.

Cebrowski, K. & Raymond, J. W. (2005). Goddard US Department of Defense, Office of Force Transformation. *Operationally responsive space—a new business model*, Report A888584.

Chen, O. & Rose, L. (2007). Boosting funding for clean energy technology. *Ubyssey*, 23 March, 16.

Choi, E. & Niculescu, S. (2006). The impact of US export controls on the Canadian space industry, *Space Policy*, *22*, 29-34.

Cirincione, J. (2000). Assessing the ballistic missile threat. Address to the Subcommittee on International Security, Proliferation and Federal Services, *Committee on Governmental Affair*, US Senate, 9 February.

Clare, L. P., Gao, J. L., Jennings, E. J. & Okino, C. (2005). Space-based multi-hop networking, *Computer Networks*, *47*, 701-724.

Cohen, W. S. (2000). Statement of the Honorable Secretary of Defense before the Senate Armed Services Committee, *Hearing on National Missile Defense*, 25 July 2000.

Collard-Wexler, S., Graham, T. J., Huntley, W., Jakhu, R., Marshall, W., Siebert, J. & Estabrooks, S. (2006). In: Space security 2006. *Waterloo*, Ontario, 10 July 2006. 146-57.

Cooper, H. F. (1996). ABM treaty costs. *Prepared testimony before the Senate Foreign Relations Committee*, 26 September.

Cooper, H. F. (2001). On spaceplanes and X vehicles. *Testimony to the House Subcommittee on Space and Aeronautics Committee on Science*, 11 October.

Cornwell, R. (2005). The real Star Wars: Bush revives missile defense plan. *The Independent*, 30 May.

Covault, C. (2007). China's ASAT test will intensify US–Chinese faceoff in space. *Aviation Week and Space Technology*, 21 January.

David, L. (2003). Military space—securing the high ground, *Space*, 2 April.

David, L. (2003b). The next shuttle: capsule or spaceplane? *Space*, 21 May.

Davis, T. M., Baker, T. L., Belchak, T. L. & Larsen, W. R. (2003). XSS-10 microsatellite flight demonstration program. In: *Proceedings of AIAA/USU conference on small satellites*, Logan, UT, USA, 11-14 August.

Day, D. (2005). Blunt arrows: the limited utility of ASATS. *The Space Review*, 6 June.

DeBlois, B. M., Garwin, B. R., Kemp, R. S. & Marwell, J. (2004). Space weapons: crossing the US Rubicon, *International Security*, *29(2)*, 50-84.

DeBlois, B. M., Garwin, R. M., Kemp, R. S. & Marwell, J. C. (2005). Star-crossed. *Spectrum*, March.

Deblois, B. M. I. (1999). Beyond the paths of heaven: the emergence of space power thought, US Air University Press, *School of Air Command and Staff College*.

Deptula, D. A. (2008). Air and space lead turning the future. *Orbis*, *52(4)*, 585-594.

DOD, (2003). Department of Defense. *Request for information for high altitude airship based relay mirror technology demonstration*. FBO Daily Issue of 10 July. Fbo #0588.

Dornheim, M. A. (2006). Lockheed Martin's secretly built airship makes first flight. *Aviation Week and Space Technology*, 5 February.

Dougherty, W., Kartha, S., Rajan, C., Lazarus, M., Bailie, A., Runkle, B. & Fencl, A. (2009). Greenhouse gas reduction benefits and costs of a large-scale transition to hydrogen in the USA. *Energy Policy, 37(1)*, 56-67.

Du, H., Hua, W., Liu, J., Gong, J., Sun, C. & Wen, L. (2005). Influence of process variables on the qualities of detonation gun sprayed WC–Co coatings, *Material Science and Engineering A, 408*, 202-210.

Duffua, S. O. & Khan, M. (2008). A general repeat inspection plan for dependent multicharacteristic critical components, *European Journal of Operational Research, 191*, 374-385.

Elhefnawy, N. (2006). The National Space Policy and space arms control, *The Space Review*, 27 November.

Elias, A. (2001). Affordable space transportation: impossible dream or near-term reality, *Air and Space Europe, 3*, 121-124.

Evans, R. L. (2007). *Fueling our future*, 1st edition, Cambridge University Press.

Fabre, H. (2002). Insurance strategies for covering risks in outer space: a French perspective, *Space Policy, 18*, 281-286.

Finarelli, P. & Pryke, I. (2006). Implementing international co-operation in space exploration, *Space Policy, 22*, 23-28.

Finney, D. (2005). Star Wars: the next generation, *the dreams and the reality*, 15 July.

Forden, G. (2006). Avoiding accidental nuclear war in south Asia: globally shared missile launch surveillance, *India Strategic, 1*, 26-28.

Freeman, D. C., Talay, T. A. & Austin, R. A. (1996). Reusable launch vehicle technology program. *International Astronautical Federation*. IAF Paper 96-V.4.01, October.

Fulghum, D. A. (2005). Microwave weapon could protect airliners, *Aviation Week and Space Technology*, 13 June.

Garamone, J. (2008). *Navy to shoot down malfunctioning satellite*. US Navy Story Number: NNS080214-19 and Navy Spotlight. USS Lake Erie website log, accessed 3 June 2008.

Garwin, R. L. (2003). Space weapons: not yet. In: Pugwash workshop on preserving the non-weaponization of space. Pugwash meeting no. 283, *Castellón de la Plàna*, Spain, 22-24 May 2003.

Garwin, R. L. (2000). Space weapons or space arms control. In: Proceedings from the American Philosophical Society annual general meeting. Symposium on ballistic missile defense, space, *and the danger of nuclear war*, 29 April.

Gates, T. S., Su, X., Abdi, F., Odegard, G. M. & Herring, H. M. (2006). Facesheet delamination of composite sandwich materials at cryogenic temperatures, *Composites Science and Technology, 66(14)*, 2423-2435.

Gloss, D., Frach, P., Gottfried, C., Klinkenberg, S., Liebig, J. S. & Hentsch, W. (2008). Multifunctional high-reflective and antireflective layer systems with easy-to-clean properties, *Thin Solid Films, 516*, 4487-4489.

Godwin, R. (2001). Rocket and Space Corporation Energia, *Apogee Books Space Series, 17*, 1st edition, 1 June.

Goh, G. M. (2004). Keeping the peace in outer space: a legal framework for the prohibition of the use of force, *Space Policy, 20*, 259-278.

Goldwasser, S. M. (2006). Information on laser technology, http://members.misty.com/don/laserioi.htm, accessed 13 January 2006.

Gordon, M. R. (1992). ‘Star Wars’ X-ray laser weapon dies as its final test is cancelled. National Desk, section A, p. 1, *column*, *5*, 21 July.

Grant, R. (2001). Is the spaceplane dead? *Air Force Magazine*, *84*, 68-72.

Grimsley, B. W., Cano, R. J., Johnston, N. J., Loos, A. C. & McMahon, W. M. (2001). Hybrid composites foe LN2 fuel tank structure, NASA Research reports, Proc. *33rd International SAMPE Technical Conference*, Seattle, W.A., U.S.A. 5-8 Nov. 1224-1235.

Gronland, T. A., Rangsten, P., Nese, M. & Lang, M. (2007). Miniaturization of components and systems for space using MEMS-technology, *Acta Astronautica*, *61*, 228-233.

Grossman, K. (1999). The phantom menace, space weapons aren’t on media radar. Extra, *Fairness & Accuracy in Reporting* (FAIR), May/June.

GS, (2006). *Globalsecurity*, HAARP, detection and imaging of underground structures using ELF/VLF radio waves www.globalsecurity.org.

GS, (2007). Globalsecurity. *Information on Aerospace Relay Mirror System (ARMS)* www.globalsecurity.org.

GS, (2008). *Globalsecurity*. Information on military spaceplane www.globalsecurity.org.

Hall, J. L. (2003). Columbia and Challenger: organizational failure at NASA, *Space Policy*, *19(4)*, 239-247.

Hammond, B. (2002). Star-crossed orbits: inside the US–Russian space alliance, *Spectrum*, *39(3)*, 57-58.

Hardesty, D. C. (2005). Space-based weapons, long-term strategic implications and alternatives, *Naval War College Review*, *58(2)*, 45-68.

Hardy, J. W. (1994). Adaptive optics, *Scientific American*, *270(6)*, 60-66.

Harr, M. E. (1989). Probabilistic estimates for multivariate analysis, *Applied Mathematical Modelling*, *13(5)*, 313-318.

Harris, A. & Harris, R. (2006). The need for air space and outer space demarcation, *Space Policy*, *22*, 3-7.

Harris, P. R. (2008). Overcoming obstacles to private enterprise in space, *Space Policy*, *24*, 124-127.

Harvey, B. (2004). China's space program — from conception to manned spaceflight. Springer Praxis Books, 1st edition, 15 July.

Harvey, B. (2007). *The rebirth of the Russian space program: 50 years after sputnik, new frontiers.* Springer Praxis Books, 1st edition, 10 May.

Hays, P. (2002). *Space Arms control and regulation: opportunities and challenges*. George Washington National Defense University, Space Policy Institute, presentation on 10 June.

Hays, P. L. (2002). United States military space into the twenty-first century. Institute for National Security occasional paper 42. *US Air Force Academy*, Report no. A770534.

Heimack, C. E. (1986). Point paper on antisatellite (asat) study. *Military uses of space*, document 00081, 27 October.

Henry, P., Compard, D., Deloffre, B., Montluc, B. D., Jamet, J. & Laffaiteur, M. (2008). The militarization and weaponization of space: towards a European space deterrent, *Space Policy*, *24*, 61-66.

Hildreth, S. A. (2005). *Ballistic missile defense*: historical overview. CRS Report for Congress RS22120, 22 April.

Hiriarta, T. & Saleh, J. H. (2010). Observations on the evolution of satellite launch volume and cyclicality in the space industry, *Space Policy*, *26(1)*, 53-60.

Hitchens, T. (2002). *Weapons in space: Silver Bullet or Russian Roulette?* The policy implications of US pursuit of space-based weapons. Presentation to the Ballistic Missile Defense and the Weaponization of Space Project Space Policy Institute, Elliott School of International Affairs, George Washington University, USA, 18 April.

Hitchens, T. (2008b). *Space Wars.* Scientific American Magazine, March.

Hitchens, T. & Chen, D. (2008). Forging a Sino-US "grand bargain" in space, *Space Policy, 24*, 128-131.

Hitchens, T., Katz-Hyman, M. & Samson, V. (2006). *Space weapons funding in the fiscal year 2007 defense budget.*

Holdstock, S. & Waterston, L. (2000). Nuclear weapons, a continuing threat to health, *The Lancet, 355*, 1544-1550.

Huntley, W. D., Bock, J. G. & Weingartner, M. (2010). Planning the unplannable: Scenarios on the future of space, *Space Policy, 26(1)*, 25-38.

Huntley, W. L. (2005). Where no bomb has gone before: US space weaponization planning and its implications. In: D. A. Ross, & S. D. McDonough, editors. The dilemmas of American strategic primacy: implications for the future of Canadian–American strategic cooperation. Toronto: *Royal Canadian Military Institute*, 18 March. 69-88.

Hyten, J. E. (2003). A sea of peace or a theater of war? Dealing with the inevitable conflict in space. In: J. M. Logsdon, & G. Adams, editors. Space weapons are they needed. Space Policy Institute, *Security Policy Studies Program*, October. 297-336.

ISB, (2004). International Security Bureau of the Department of Foreign Affairs and International Trade Canada. Space Security 2003. *Ottawa*, Canada, March 2004.

Jasani, B. (1985). Space weapons, *Space Policy, 1*, 164-178.

Jasani, B. (2001). US national missile defence and international security: blessing or blight? *Space Policy, 17*, 243-247.

Johnson-Freese, J. & Erickson, A. S. (2006). The emerging China–EU space partnership: a geotechnological balancer, *Space Policy, 22*, 12-22.

Jones, C. (2006). The axis of non-proliferation, *Problems of post-communism, 53(2)*, 3-16.

Jonson, N. (2003). Shuttle disaster will renew interest in replacement program. Aviation Week 2003, *Aerospace Daily & Defense Report*, 4 February.

Kammen, D. M. (2006). The rise of renewable energy, *Scientific American, 295*, 84-93.

Kingwell, J. (1990). The militarization of space: a policy out of step with world events? *Space Policy, 6*, 107-111.

Kleiman, P. (2006). Unique mission focuses on exposing systems vulnerabilities. *AFRL Space Vehicles Directorate*, Public Release.

Koert, P. (1990). *Artificial ionospheric mirror composed of a plasma layer which can be tilted.* US Patent no. 5, 041,384, APTI Inc., Washington, DC, USA, 17 May.

Kornilov, Y. P. (1992). Space Program: The little-known Polyus, Zemlya i Vselennaya, no. 4, July–August, 18-23. *Translated in JPRS Report, Science & Technology*, Central Eurasia: Space, 25 March 1993 (JPRS-USP-93-001), 23-4.

Kramer, H. J. (2001), *Observation of the earth and its environment*: survey of missions and sensors (4th edition), Springer, New York.

Kries, W. V. (2002). The demise of the ABM treaty and the militarization of outer space, *Space Policy, 18*, 175-178.

Lebeau, A. (2008). Space: The routes of the future, *Space Policy, 24*, 42-47.

Lewis, J. (2004). Autonomous proximity operations: a *coming collision in orbit*? www.armscontrolwonk.com, 11 March.

Lewis, J. (2006). Information on US military spending www.armscontrolwonk.com, accessed 29 March.

Lewis, R., Kennedy, M., Ghashghai, E. & Bitko, G. (2005). Building a multinational global navigation satellite system: an initial look. *Monograph MG-284-AF*, Rand Corporation, Santa Monica, USA.

Liao, S. H. (2005). Will China become a military space superpower? *Space Policy, 21(3)*, 205-212.

Llanes, L., Torres, Y. & Anglada, M. (2002). On the fatigue crack growth behavior of WC–Co cemented carbides: kinetics description, microstructural effects and fatigue sensitivity, *Acta Materialia, 50*, 2381-2393.

Lloyd, R. (1999). *NASA to develop intelligent, cake-sized satellites* CNN News, 20 August.

Logsdon, J. M. (1998). Commercializing the international space station: current US thinking, *Space Policy, 14*, 239-246.

Logsdon, J. M. (2008b). Why space exploration should be a global project, *Space Policy, 24*, 3-5.

Logsdon, P. N. (2008a). Reinvigorating transatlantic space relations: the joint ESPI–SPI memorandum, *Space Policy, 24*, 119-123.

Longwell, J. P., Rubin, E. S. & Wilson, J. (1995). Coal: energy of the future, *Progress in Energy and Combustion Science, 21(4)*, 269-360.

Lovell, A. C. B., Blackwell, M., Ryle, D. E. & Wilson, R. (1962). West ford project, interference to astronomy from belts of orbiting dipoles (needles), *Journal of the Royal Astronomical Society, 3*, 100-109.

Malenfant, P. R. L., Lee, J. U., Li, Y. & Cicha, W. V. (2006). High performance field effect transistors comprising carbon nanotubes fabricated using solution based processing. *US Patent Application*, 20060081882, 20 April.

Mendillo, M., Hawkins, G. S. & Klobuchar, J. L. (1975). A sudden vanishing of the ionospheric F region due to launch of Skylab, *Journal of Geophysical Research, 80(60)*, 2217-2228.

Milas, V. F., Vouyioukas, D., Moraitis, N. & Constantinou, P. (2008). Spectrum planning and performance evaluation between heterogeneous satellite networks, *European Journal of Operational Research, 191*, 1132-1138.

Moltz, J. C. (2007). Protecting safe access to space. Lessons from the first 50 years of security, *Space Policy, 23*, 199-205.

Montluc, B. D. (2010). Russia's resurgence: Prospects for space policy and international cooperation, *Space Policy, 26(1)*, 15-24.

Morris, J. (2005). Air Force backs away from near-space maneuvering vehicle. *Aerospace Daily and Defense Report*, 18 March.

Nath, C. (2006). Military uses of space. *UK Parliamentary Office of Science and Technology*, POSTnote 273, 1-4.

Ning, X. & Fang, J. (2007). An autonomous celestial navigation method for LEO satellite based on unscented Kalman filter and information fusion, *Aerospace Science and Technology, 11*, 222-228.

Novosti, R. (2008). China's space ambitions. *India Strategic*, August.

NP, (2005). National Priorities. *Information on National Security* www.nationalpriorities.org, 15 August 2005, accessed 19 February 2006.

O'Shaughnessy, N. J. (2004). Politics and propaganda—weapons of mass seduction, University of Michigan Press, Michigan, USA.

Oberg, J. (2005). *Air Force microsatellite passes key first tests XSS-11 successfully completes series of orbital rendezvous maneuvers*, MSNBC 9 September.

Oberg, J. (2006). Space weapons: hardware, paperware, beware? *The Space Review*, 13 November.

Oberg, J. (2007). The dozen space weapons myths, *The Space Review*, 12 March.

Oberg, J. (2008). US satellite shootdown: the inside story. *Institute of Electrical and Electronics Engineers*. Spectrum online, article 6533, August.

ONR, (2004). Office of Naval Research. Contract no. M67854-04-C-5074, Atlanta, USA, 1 July.

Page, J. T. (2006). Stealing Zeus's Thunder, Physical Space-Control Advantages against Hostile Satellites, *Air & Space Power Journal*, Summer, 1 June.

Parry, D. (2004). *MD Robotics quoted in Precarn*. Autonomous space robotics—a technological and commercial success story www.precarn.ca, November.

Partch, R. E., Baker, V. & Keith, C. (2003). *Autonomous proximity microsatellites* www.afrlhorizons.com, December.

Petronia, G. (2010). Space technology transfer policies: Learning from scientific satellite case studies, *Space Policy, 26(1)*, 39-52.

Phillips, T. & O'Leary, B. (2003). Common aero vehicle on orbit. Schafer Corporation, Report to the Defense Technical Information Center, *Manpower and Training Research Information System*, 6 September.

Pike, J., Vick, C. P., Berman, S. D., Lindborg, C. & Sherman, R. (1999). Military Spaceplane X-40 space maneuver vehicle integrated tech testbed, FAS Space Policy Project, *Military Spaceplane Project*.

Pincus, W. (2005). Pentagon has far-reaching defense spacecraft in works, *Bush administration looking to space to fight threats*. Washington Post A03; 16 March.

Possel, W. H. (1998). Lasers and missile defense: new concepts for space-based and ground-based laser weapons. Air War College, *Center for Strategy and Technology*, Air University, Alabama, Occasional paper no. 5, July.

Postol, T. A. (1992). *Optical evidence indicating patriot high miss rates during the gulf war.* Testimony to the Committee on Government Operations, US House of Representatives, 7 April.

Preston, R., Johnson, D. J., Edwards, S. J. A., Miller, M. D. & Shipbaugh, C. (2003). Space weapons Earth wars. Monograph MR-1209, *Rand Corporation*, Santa Monica.

Qingguo, J. (2005). One administration, two voices: US–China policy during Bush's first term, *International Relations of the Asia-Pacific, 6(1)*, 23-36.

Raffaelle, R. P., Landi, B. J., Harris, J. D., Bailey, S. G. & Hepp, A. F. (2005). Carbon nanotubes for power applications, *Materials Science and Engineering B, 116*, 233-243.

Randow, G. V. (2005). Neue Saebel in Fernost (New sabers in the far east), *Die Zeit*, 21 April, 17.

Rosati, G., Boschetti, G., Biondi, A. & Rossi, A. (2009). Real-time defect detection on highly reflective curved surfaces. *Optical Laser Engineering, 47(3-4)*, 379-384.

Rose, L. (2006). Spin coated $Ce_{1-x}Gd_xO_{2-d}$ layers for anode supported fuel cells, *Tidsskrift for Dansk Keramisk Selskab*, *1*, 16.

Rose, L. (2010). Review of cornerstone parameters influencing future energy policy, Chapter 10, in: Global Environmental Policies: Impact, Management and Effects, Eds. R. Cancilla, & M. Gargano, *Nova Environmental Research Advances Series*, Hauppauge, NY, U.S.A.

Sadeh, E. (2009). Space policy challenges facing the Barack Obama administration, *Space Policy*, *25(2)*, 109-116.

Samson, V. (2008). The political costs of shooting down USA-193. *Center for Defense Information/World Security Institute*, 27 February.

Schaffer, A. M. (2008). What do nations want from international collaboration for space exploration? *Space Policy*, *24*, 95-103.

Scully, M. O. (1992). From lasers and masers to phaseonium and phasers, *Physics Reports*, *219(3-6)*, 191-201.

SD, (2004). *US could shoot down Euro GPS satellites if used by China in wartime Spacedaily*, 24 October 2004.

Secara, T. & Bruston, J. (2010). Current barriers and factors of success in the diffusion of satellite services in Europe, *Space Policy*, *25(4)*, 209-217.

Sega, R. M. & Cartwright, J. E. (2007). Plan for operationally responsive space. A report to Congressional Defense Committees, *Department of Defense*, 17 April.

Shachtman, N. (2004). Airborne laser in trouble again, *Defensetech*, 5 January.

Singer, J. & Bates, J. (2005). Critic differ on XSS-11 mission objective. *Space News* www.space4peace.org, 18 April.

Skotnikov, L. A. (2002). Possible elements for a future international legal agreement on the prevention of the deployment of weapons in outer space, the threat or use of force against outer space objects. In: Conference on disarmament, 27-28 June, *Foreign Ministry Document*, 1324-28-06-2002.

Smith, B. (1999). The challenge of space power, *Airpower Journal*, *13*, 32-39.

Sojka, J. J., Eccles, J. V., Thieman, H., Sridharan, R., Lakhina, G. S. & Rao, P. B. (2002). An observation-driven model of the equatorial ionosphere — DEOS rocket campaign study, *Advances in Space Research*, *29*, 899-905.

Spacy, W. L. (1999). Does the United States need space-based weapons? College Of Aerospace Doctrine, Research and Education. Air University, Cadre paper. Alabama: Air University Press, *Maxwell Air Force Base*, 36112-6610, September.

Stadd, C. & Bingham, J. (2004). The US civil space sector: alternate futures, *Space Policy*, *20*, 241-252.

Steigies, C. T., Block, D., Hirt, M., Piel, A. & Thiemann, H. (2000). Electron density measurements with impedance and Langmuir probes in the DEOS campaign, *Advances in Space Research*, *25*, 109-112.

Steigies, C. T., Hirt, M. & Pie, A. (2002). Electron density and temperature measurements obtained in the DEOS campaign, *Advances in Space Research*, *29*, 893-898.

Thiemann, H., Sojka, J. J., Eccles, J. J., Rao, P. B., Rao, P. V. S. R. & Sridharan, R. (2001). Indo-German low-latitude project DEOS: plasma bubbles in the post sunset and nighttime sector, *Advances in Space Research*, *27*, 1065-1069.

Tillinghast, T. & Katzman, J. (2005). SBIRS dilemma: go high or go home. *Defense Industry Daily*, 15 April.

Tirpak, J. A. (2004). Air Force plans call for defensive and offensive systems to protect vital US spacecraft. Securing the space arena, *Air Force Magazine*, *87(7)*.

Toso, A. R. (2004). Systems-level feasibility analysis of a microsatellite rendezvous with non-cooperative targets, Report afit/gss/eny/04-m06, Department of the air force, *Air Force Institute of Technology*, Wright-Patterson Air Force Base, Ohio.

UN, (2008). UN Office Geneva. Conference on disarmament hears address by Foreign Minister of Russia and message by Chinese Minister For Foreign Affairs, *Foreign Minister Lavrov presents Joint Russian–Chinese draft treaty to prevent the placement of weapons in outer space*, Sciptum of statements made regarding Presidential Draft Decision CD/2007/L.1, 12 February 2008.

USAF, (2003). Headquarters USAF/XPXC. *The US air force transformation flight plan*, November 2003.

USDS, (1967). US Department of State, Bureau of Arms Control. Treaty on principles governing the activities of states in the exploration and use of outer space, *including the moon and other celestial bodies*, 10 October.

USOTA, (1995). US Office of Technology Assessment, The US national space transportation policy: issues for congress, *Space Policy*, *11*, 283-287.

Varni, J. G. G., Powers, G. M., Crawford, D. S., Jordan, C. E. & Kendall, D. L. (1996). Space operations-through the looking glass, global area strike system. *Air Force*, 2025, vol. 3, August [Chapter 14].

Waldrop, E. (2004). Weaponization of outer space: US policy, *Annals of Air and Space Law*, *29*, 1-28.

Walter, K. (2002). Faster inspection of laser coatings, science and technology review, *Photothermal Microscopy*, March, 17-19.

Ware, R. H., Rogers, T. F., Padua, D. J. & Roberts, W. O. (1988). Space Phoenix, *Space Policy*, *4*, 143-150.

Weiner, T. (2005). Air Force Seeks Bush's Approval for Space Weapons Programs. *New York Times*, Section A, 1, column 6, 18 May.

WH, (2003). White House Press Release, *National strategy for combating terrorism*, Office of the Press Secretary.

Wiedemann, C., Bendisch, J., Krag, H., Wegener, P. & Rex, D. (2001). Modeling of copper needle clusters from the West Ford Dipole experiments. In: Huguette Sawaya-Lacoste, editor. Proceedings of the third European conference on space debris, 19-21 March 2001, *European Space Agency*, Darmstadt, Germany, SP-473, vol. 1, 315-20.

Wilson, J. (2002). Return of the battle blimps. *Popular Mechanics*, 12 March.

Wright, D., Grego, L. & Gronlund, L. (2005). The physics of space security, *American Academy of Arts and Sciences*, Cambridge, USA.

Wu, G. M. & You, Z. D. (2007). Enhanced light extraction efficiency by surface lattice arrays using blue InGaN/GaN multiple quantum wells, *Solid-State Electronics*, *51*, 1351-1359.

Young, K. (2005), Autonomous military satellite to inspect others in orbit, *New Scientist*, *12*, April.

Zhou, Y. (2008). Perspectives on Sino-US cooperation in civil space programs, *Space Policy*, *24*, 132-139.

Zielinski, R. H., Worley R. M., Black D. S., Henderson S. A. & Johnson, D. C. (1996). Star trek—exploiting the final frontier-counterspace operations in 2025. *Air Force*, 2025, *vol. 3*, August, Chapter 9.

Zimmerman, P. (1990). Remote sensing, super power relations and public diplomacy, *Space Policy*, *6*, 19-32.

Zimmerman, P. (1991). The role of satellite remote sensing in disaster relief, *Space Communications*, *8*, 141-152.

Zraket, C. A. (1992). *Before the legislation and national security subcommittee of the Committee on Government Operations*, US House of Representatives, 7 April.

In: Space Policy and Its Ramifications
Editor: John P. Ramos

ISBN: 978-1-61761-555-9

Chapter 4

DIVERGING 21ST CENTURY GLOBAL SPACE POLICY GOALS: THE MERCHANTS, THE GUARDIANS AND THE CIVIL GOVERNMENT ADVOCATES[1]

Joseph N. Pelton
Emeritus, Space and Advanced Communications Research Institute,
George Washington University, Washington, DC,
and Former Dean, International Space University

INTRODUCTION

It seems almost shocking to those who remember first hand the launch of Sputnik in October 1957 that the first fifty years of human space activities are now complete and then some. Certainly, more than a half-century of space-related experience represents enough time for an honest and thoroughgoing critical analysis of what we have now accomplished in space exploration, applications and science. Perhaps it is even more important to take this longer-term perspective to assess honestly where, why and how we have failed. This is a critical step toward assessing what future goals should be set for space in the coming half century. Such an exercise should let us consider who should be in charge of setting our future objectives in terms of space policy and a viable regulatory framework going forward. This should particularly focus on whether international, regional or national actors will be in charge of implementing and enforcing these space-related policies and regulations. It should also address what appears to be divergent-- as opposed to convergent—space policy perspectives as held by the various civil space agencies, the military or defense-related agencies, and the

[1] This article was, in part, intellectually stimulated by an article first prepared over a decade ago by Scott Pace , now director of the Space Policy Institute, at George Washington University. See: Scott Pace, "Guardians and Merchants: Balancing National Interests in Space Development, International Space Policy Forum, (1999) George Washington University, Elliott School of International Affairs, pp. 5-66 (This article which actually addresses U.S. space policy dilemmas stemming from "dual use" of civilian space assets is available through the RAND reprint series).

growing number of commercial entities? These three groups of critical actors in the space arena seem, in many instances, to have quite different views on how to best proceed both in terms of implementing space programs and activities and what are the best forms of space policy and regulation.

Key questions include the following: Should quite expensive human space exploration be our primary goal in either a global or national context? Should practical space applications—related to communications, remote sensing, climate change, meteorology, space navigation, disaster assessment and recovery and solar power-- represent a priority goal over human space exploration? Should deep space missions be essentially restricted to robotic systems and sensors to carry out space science? Should commercial space ventures be encouraged to thrive and supplant civil governmental space activities wherever and whenever possible—or at least in near Earth orbit? What about the strategic uses of space for military and defense related purposes? Should there be limits to the militarization of space and what form of collaboration, cooperation or formal divisions should there be with regard to commercial, civil governmental and military space operations? Who should be in charge of space safety regulations and their implementation?

These and a wide range of other questions seem ripe for serious analysis, study and future goal- setting purposes. In many ways these questions, for the next half century, may well come down to who leads and who follows? Will the leaders in space activities and space policy formation be the "Merchants" (the commercial space entrepreneurs), the "Guardians" (those that utilize space systems for military and defense-related purposes), or the "Civil Space Advocates & Regulators" (the official national and regional space agencies and regulatory authorities)? Which of these will become the dominant force in space—especially from Clarke Orbit (geosynchronous orbit) on down? Which of these "actors", working in tandem (or not) with the United Nations and other international agencies, will ultimately decide how we prioritize our outer space activities. In short, going forward, who will set space policies and regulations and who will "enforce" them? This, of course, is a lot for one article to try to address.

Thus the purpose of this article is to focus on one particular aspect of these longer-term outer space-related issues. This is to analyze the often different perspectives of the so-called "merchants", "guardians" and "civil space advocates and regulators" when it comes to space policy and regulation in some specific programmatic areas.

In broad brush perspective, many of the "merchants" as reflected by such groups as the "Commercial Spaceflight Federation" is for an open and "laissez faire" approach to the regulation of outer space in and around Earth. This perspective is in some ways like the "Law of the Seas" approach which suggests that once one is beyond national airspace limits (i.e. the equivalent to and above national territorial waters adjacent to a nation's landmass) there is a minimum of regulation beyond basic "rules of the road". This "open approach" would have every national or commercial entity taking care of their own projects' safety. The "guardians" have accepted the current broad and vague U.N. Outer Space Treaty (formally known as the *Treaty on Principles Governing the Activities of States in the Exploration and Use of Outer Space, including the Moon and Other Celestial Bodies)* This treaty indicates that there should not be "militarization of space", but leaves the precise definition of what militarization means wide open. Article IV does not explicitly ban "manned military spacecraft", although it does specify that no nuclear weapons are to be launched into orbit. Certainly within the current wording of the Outer Space Treaty military entities have deployed dedicated and dual-use communication satellites for military purposes as well as navigational satellites for missile targeting and Earth observation satellites for surveillance and intelligence gathering systems. Sub-orbital regional and intercontinental missile systems carrying weapons of mass destruction still exist and planning and engineering for so-called "star wars-like" systems for defense and "potentially offensive" purposes have been undertaken. When the UN General Assembly took a vote in 2008 on a measure entitled "Prevention of An Arms Race in Outer Space," 166 nations voted in favor of this resolution with two abstentions and the U.S. being just the one lone nation voting against the resolution.

Finally there are the civil space agencies of the world. These agencies are in synch with national governments and international agencies such as the United Nations Committee on the Peaceful Uses of Outer Space, the International Telecommunication Union, etc. The major space agencies have generally worked toward an internationally harmonized approach to outer space. This has been seen in close cooperation around the International Space Station (ISS), the new "Global Exploration Strategy" (in progress since 2004) that spells out the rationales for carrying out a long-term program of human space exploration, and cooperation within the United Nations to develop voluntary principles to limit the growth of orbital space debris.

In short, many of the "space merchants" would like to see a wide-open, entrepreneurial approach to space and space regulation. The "space guardians" would like to see a minimum of new military or defense-related restrictions on what can be done in outer space and yet have serious concerns about what private space entrepreneurs might be allowed to do in terms of operating private space planes or private space stations. The "civil space agencies" generally see the need for more international cooperation in space, common rules, regulations and standards for space safety, and an improved regulatory framework for cooperative space ventures.

The analysis that follows examines several emerging future space-related programs and activities. The focus of the analysis is how the "merchants", "the guardians" and "civil space advocates" might wish these activities to be regulated in terms of strict controls or much looser and free-form guidelines.

The Future of Space: New Directions for the Next Fifty Years

Space agencies, particularly those of the U.S. and the former Soviet Union, have spent the bulk of their money over the past half-century on human space exploration. Nevertheless NASA and RSA (Roscosmos) have supported not only "manned space initiatives" but also, to varying degrees, strongly funded space science and applications as well. With the exception of NASA and Roscosmos, the other major space agencies (i.e. plus now the European Space Agency, JAXA, ISRO and the Chinese Space Agency) have focused largely on what might be called "mission to planet Earth" initiatives in terms of space applications and the space sciences. In recent years, however, China and even ESA and JAXA, have increased their "human space exploration" budgets.

Military officials around the world have, for the most part, had their own strategic space agendas. Certainly they have often worked in tandem with government space agencies and aerospace companies, but their main focus has been on the development of missile systems such as intercontinental ballistic missiles (ICBMs) and space defense technology such as the Strategic Defense Initiative (SDI), popularly known as "Star Wars". Recently defense spending has increased in areas related to space plane technology—for both manned and non-manned versions. This is not to say that the governmentally funded civil and defense space programs have not often transferred technology across the often hard to define line between military and non-military programs. Civilian and military space programs, going forward, are more likely to have divergent rather than common objectives.

Commercial applications of space by corporate enterprises were, for decades, quite limited. Serious profitable enterprises beyond communications satellites were hard to define and even harder to bring to market with viable profitability. Nevertheless enterprises related to space navigation and remote sensing have now become billion dollar businesses. Serious efforts to commercialize space travel and to create commercial solar power satellite systems are now underway. Virgin Galactic has indicated that they will start flights for paying customers in 2011 and the new private U.S. venture seeking to commercialize solar power systems, Solaren, has contracted with electrical power companies to begin delivering commercially derived solar power satellite electrical energy as of 2016.

As all three of these various types of space activities have continued to evolve, i.e. the civil governmental space programs, the military space programs, and the commercial space businesses, the regulatory environment for the use of outer space has also begun to mature. This has led to an increasing number of international regulatory provisions (in terms of space treaties, international agreements, and U.N. backed guidelines and best practices). We have also seen at the national, regional and international level various types of space policies and standards come into being. Some of these have been based on practices and precedent related to international waters and the open seas, others related to international air traffic regulation, and yet others that have evolved *de novo* from international discussions and scientific and political exchanges.

Within the United Nations a series of Space Treaties, Conventions and Agreements have come into being and helped to set a least some basic "rules of the road" for space activities.[2] In the first few years (i.e. the late 1950s and the 1960s) the struggle around the world was simply to be able to launch payload reliably—any payload that made it successfully into orbit in the early years was a success. But today—a half century later—tough questions are being asked about the purpose of space programs. The cohesion that bound together the civil space programs, military space programs and commercial space programs is no longer as evident as it was decades earlier. The defense and civil space programs are today at the national or regional level sometimes vying for the same limited governmental funds. The new commercial space entrepreneurs have sometimes—even often--expressed that the conventional, highly bureaucratic approach to space taken by an ESA or NASA is slow, expensive, and "poorly managed" when compared to new start-up commercial space ventures. Outspoken entrepreneurs such as Burt Rutan of Scaled Composites, Elon Musk of Space X, Jeff Gleason of X-Cor and even Sir Richard Branson of Virgin Galactic have extolled the virtues of a new commercial pathway to more flexible, innovative and "less stuffy" approaches to new space ventures. They suggest that in the area of commercial space

[2] Ram Jakhu, John Logsdon, and Joseph N. Pelton, "Space Policy, Law and Security" (Chapter 9) in Joseph N. Pelton and Angelina Bukley, *The Farthest Shore: A 21st Century Guide to Space*, (2010) Apogee Press, Canada,

travel the "extension" of the safety practices and standards developed by the civil space agencies would essentially be a mistake. They view the approach of the civil space agencies as being overly "bureaucratic, time consuming, and extremely labor intensive". They have argued that the best approach is to develop entirely new safety practices and standards, born of innovative new technology.

Space Policy and Regulations Formation and Implementation for the Future

The next fifty years of outer space activities will likely be much more challenging in terms of global space policy being able to transcend the increasingly different interests and objectives of the key players discussed above. These "actors" in the field of space and the "definers" of space policy in future decades not only see outer space activities in fundamentally different ways but have different objectives, thought processes and certainly different priorities.

In short, new technological, political, defense and economic opportunities may well lead to increasing conflicts over best forms of management practices, appropriate regulations and standards for oversight, coordination and safety, best ways to finance and implement new space initiatives, and even the best way to establish priorities for new projects and initiatives. Any comprehensive attempt to create a harmonious and systematically agreed set of standards and space policies is likely to run afoul of the interests and basic mindsets of the Merchants, the Guardians and the Civil Space Advocates.

Certainly there will be some limited areas of global space policy where common ground can be found. There is likely to be agreement on the need to combat climate change, to protect the Ozone layer, to reduce orbital debris, and to adopt basic regulations and standards related to improved space safety where simple common sense and understanding can be applied. But even in these areas there are likely be disagreements as to which are the best technological approaches to use. Likewise there may well be differing viewpoints as to whether commercial, military or governmental space agencies should be taking the lead when it comes to implementation.

Certainly there will likely be conflict between and among the merchants, the guardians and the space bureaucrats, even as to "who should set space policy and whether the defining regulatory process should be at the national, regional or global level?"

There is very likely to be disagreement as to who can exploit space and for what purpose and where the line between commercial space and governmental space projects should be drawn. Likewise there is clearly disagreement as to who or what agency, and with what authority, should be the legitimate enforcers of future space policy and regulation? The recent evolution of new commercial space ventures seeking to offer "space adventure rides" for paying passengers seeking to go on a two hour ride just above the stratosphere to experience dark sky and five minutes of weightlessness has raised a whole host of new space policy and regulatory questions.

Those like Robert Bigelow who wants to operate a private low orbit space station and offer in-orbit stays to paying guests know that such commercial ventures in space pose a whole host of new and unanswered questions in the arena of what might be called strategic space policies. How should such commercial space stations be controlled? Do decisions related to commercial space stations create new precedents that might carry over into such areas as commercial space colonies in outer space or on the Moon or other celestial bodies. Regulatory policy with regard to other commercial projects in space such as solar power satellites (SPS) also arise. In this arena, there could be issues as to whether such commercial systems could be used as weapons systems, whether SPS operations might pose health or safety risks to aircraft passengers or might through reflected energy endanger remote sensing or particularly telecommunications satellite operations. If we are to think truly longer term then the regulatory processes that are developed for outer space in the next decade or so could set key precedents that might affect the future opportunities related to mining of asteroids, terra-forming of a planet or more simply the future direction of commercial space transportation systems.

Should the approach to future space regulations and policies focus on commercial incentives, opportunities and even possible revenue streams through taxes? Or should there be penalties for not adhering to some clear-cut rules with a focus on governmental or even intergovernmental controls? Or will there be one set of policies, standards, regulations and sanctions that apply to private space ventures and another that applies to national governments or military operations? And if so who or what will have the power to consider, enact and especially to enforce global space policies and regulations?

In Europe, there is a thought process that seems to be gravitating toward the idea that many of the issues are similar to regulating and setting safety standards for air traffic control. Using this line of international law and regulatory process, there is sentiment that the European Aviation Safety Agency (EASA) should set the rules and enforce them—or perhaps eventually the U. N. Agency, the International Civil Aviation Organization (ICAO), should be the regulatory oversight body. In contrast, in the United States the Aviation Safety Advisory Panel from the commercial world is essentially telling the U.S. FAA and the U.S. Congress that there is no world space safety police or global space regulators and that there should be none. They argue that outer space is actually like the high seas and other than a few common sense rules, there will be no need to regulate outer space for some time to come. This is, however, but one example where the regulatory process related outer space diverges between and among the various actors—both geographically and by functional interest.

In this article some of the more important issues to be faced in this regard are explored along with possible mechanisms to reduce disagreements among the key actors and to allow more effective space policy to evolve. Particular areas to be addressed include: (i) commercial space travel and exploration; (ii) global regulation of and improved safety standards for air and space travel safety; (iii) solar power generation; (iv) militarization of space; and (v) for the longer term, the mining of asteroids, the establishment of space colonies and even the terra-forming of planets. Although there are a host of other areas that might be considered, these five areas of future global space policy clearly represent areas of potential conflict. In many cases there will be disagreement between and among the merchants, the guardians and the space bureaucrats, but also there will be disagreements between various geographic sectors of the world. Today, these perspectives on necessary space polices and their enforcement are primarily differences between the U.S. on one hand and Europe on the other, but in coming years, Asian perspectives—primarily from Japan, China and India—will become increasingly apparent and important.

It is useful to consider where we may be headed in terms of outer space activities for the next fifty years. In particular, it is important to analyze current differences in viewpoint among the various actors to see where future clashes as to regulatory approaches might be found and where possible accommodation might be achieved. This analysis will recall ways in the past where useful processes have been invoked to help sort out differences. Further this process may also help identify what some of the thorniest challenges might be in terms of arriving at new regulatory agreements, and in terms of uncovering basic differences of opinion between and among the key "space actors" or regional groupings who either accept (i.e. Europe and Canada) or disagree (i.e. the U.S.) with more regional or global regulatory process and enforcement powers.

Future Space Regulations and Commercial Space Travel and Exploration

There have been commercial space transportation launches for some time. The Intelsat global satellite system, now Intelsat Ltd., began launching commercial communications satellites on Delta launch vehicles in 1965. During the Reagan Presidency in the United States, the U.S. Government created an Office of Commercial Space Transportation within the U.S. Department of Transportation under the Federal Aviation Administration (FAA-AST) to regulate U.S. commercial launches and oversee their safety. This transfer of regulatory oversight from NASA to the FAA was designed to spur the growth of these launch services on a commercial basis. Today there is a wide range of commercial launch services available around the world. The Arianespace commercial consortium leads this commercial market with its Ariane 5 launcher. In addition commercial launch services are available from the United States (Atlas, Delta, Taurus, Pegasus, etc.) from the Ukraine (Zenit), India (the Polar Satellite Launch Vehicle), China (Long March) and a variety of Russian launchers as well as international consortia such as International Launch Services (ILS) and Sea Launch. These commercial launch services launch various payloads and do not carry human passengers. U.S., Russian and Chinese launches with crew aboard are today strictly governmental operations.

There is a range of options whereby commercial launches with human crew and/or passengers aboard may take place in future years. These include not only the sub-orbital flights of space planes and other commercial "space adventure" transport systems, but there are also more ambitious efforts underway. NASA is pursuing the so-called Commercial Orbital Transportation Systems (COTS) where in the future contractors might fly astronauts to and from the International Space Station (ISS).[3] Bigelow Aerospace Corporation and InterOrbital Systems (IOS) are both involved in plans to operate a low Earth orbit (Leo) commercial space station. Bigelow is sponsoring the so-called America's Space Prize to spur the development of commercial transport to its inflatable private space station.[4] IOS is developing its Neptune TSAAHTO (Two Stage and a Half to Orbit) vehicle with a 5 person orbital module. This vehicle could also possibly win the $50 million America's Space Prize and perhaps even more lucrative, lead to future contracts to ferry passengers to the Bigelow space station.[5]

---

[3] Joseph N. Pelton and Peter Marshall, License to Orbit: The Future of Commercial Space Travel (2009) Apogee Press, Canada

[4] Bigelow Aerospace private space station initiatives, Genesis II and the America's Space Prize. See www.bigelowaerospace.com/genesis_II/

[5] Inter Orbital Systems (IOS) Launch Development Program, www.interorbital.com

After years of space plane research, NASA has now abandoned the field of space plane development to the U.S. Air Force and to private entities such Scaled Composites, the Space Ship Corporation, XCOR, etc. Virgin Galactic and XCOR now plan to start commercial "experimental flights" in 2011.

This array of future possible commercial space transportation launch services and private space stations involving the safety transport of private citizens or astronauts or possibly military personnel raise a host of issues. These issues sort out into: (i) **Safety issues.** Here the safety is not only for the crew and passengers on board space planes or private space stations but also the safety of people on the ground and the safety of people flying in airspace who might be endangered; (ii) **Strategic issues.** These issues concern who is "licensed" to operate commercial vehicles or private space stations and what controls, incentives, and/or potential penalties are exercised by national governments or supranational agencies to ensure that these commercial vehicles or stations are not used for military purposes; (iii) **International Regulatory and Trade Issues**. There are a host of ancillary issues here such as: (a) licensing and control of national or supranational launch sites, (b) the assigning and allotment of radio frequencies for communications with the launch vehicles or private space station, (c) responsibilities related to air traffic control during ascent and descent (local national airspace and ICAO regulations), and (d) the responsibility for and jurisdiction of entities involved in the taxation of such commercial operations as well as the proper application of national and international trade provisions (i.e. within the country of launch and landing and within applicable World Trade Organization WTO provisions.)

Figure 1. The X-37, an Early NASA Development, Preceding Current Commercial Space Planes (Graphic Courtesy of NASA)

Differing Perspectives on Regulating Commercial Space Transportation & Space Stations

Key Actors	Safety Regs. & Standards	Security Concerns	Air Traffic Control /Harmonization	Other Regulatory/ Trade/Tax/Environ. Issues
Civil Space Agencies	Would like to see consistent safety standards based on space agencies experience.	Would like to secure space agency assets. Such as enforce strict 1 km "bubble around ISS.	Aircraft Control Agencies foresee increased dangers as both air, hypersonic air and commercial space traffic increases. Envision need for integrated controls.	**Tax issues:** Such as appropriate VAT; **Trade issues:** Such as whether commercial space traffic will morph into hypersonic transport? **Radio Frequencies Stand:** Need for interoperable air to commercial space comm.? **Environ:** Restrictions on rocket emissions re. Ozone layer
Military	Would also like to see consistent standards based on previous military and space agency experience.	Would like to governmental control over all space flights and not see private space stations/colonies	Also foresees conflict between commercial air, military air, UAV and commercial space flight control. Would like to see strict controls & military priority.	Only major concerns would be point to point hypersonic flight and RF frequency coordination/interference. Would likely resist environmental controls on launcher emissions.
Commerical	Would like less government involvement. Freedom to develop new standards/new technology. Branch out from past govt/mil. Space standards	Would like to be free of military controls. Concerns about further militari-zation of space hurting their market expansion.	Again would like minimum controls with regard to airspace and particularly resist ICAO or other agency assuming "space traffic control".	**Taxes:** Seeking minimum tax liability. Seek ability to operate from low tax spaceports. **Trade:** Seek minimum trade constraints. Global reciprocity. **Radio Frequencies:** Support need for interoperability; **Environment:** Resist controls. Dirtiest solid fuel "throttle-able" rockets may be safest.
International Agencies (ICAO, ITU, WTO, UN)	All UN agencies foresee greater role: ICAO-air traffic control. ITU-freq. coord.; WTO-specific inclusion in GATS agree., UN-concerns about radiation, pollution, orbital debris.	Foreseen need for additional UN based treaties covering rights & obligations for space transport operators and especially for private space stations or colonies. Restriction on military uses.	ICAO for space, European Aviation Safety Agency (EASA) likely to assume additional responsibilities despite US government position of limited international traffic control of space transportation. Europe, Canada and others have opposite view to U.S.	UN and various specialized agencies will develop additional controls and responsibilities in relevant areas related to air traffic control/safety, RF freq. allotments, trade and tax provisions and orbital debris and pollution controls.

Solar Power Generation

Space applications, starting with space telecommunications and growing to include remote sensing, Earth observation, space navigation and meteorology, have largely not posed major international policy or regulatory issues requiring specific action within the UN or other international processes. The development of UN guidelines regarding the use of nuclear power does, however, stand out as a clear exception. Remote sensing, Earth observation and surveillance satellites and the use of space navigation for missile targeting, and direct broadcast satellite transmissions spilling over national boundaries have, of course, all given rise to international disagreements and sharp differences within the United Nations at one time or another, but the end of the Cold War and wide-spread use of these space applications technologies have largely dissipated these issues as major international concerns with regard to the use of outer space. The increase of orbital debris has also been an issue that all operators of application satellites have given increased attention to in light of the danger this risk now poses. These operators are seeking to lessen this danger and are actively supporting the UN guidelines now adopted on this subject.

The increasingly likely deployment of solar power satellites gives rise to a series of new concerns that will very likely give rise to calls for new standards, regulations and policy guidelines for this new type of commercial space service. Gary Spirnak, CEO of the Solaren Corporation, has work with a team of former Hughes Corporation engineers to design what he believes is a commercially viable design for a workable solar power satellite that can beam power back to Earth 24 hours a day. Solaren has actually signed a contract with a start in delivering power to a U.S. electric utility company as of 2016. This SPS, as currently designed, would include three separate components--with each having a different function. The first part would a very large and lightweight mirror that collects and focuses sunlight on the second component that would be a grid of the solar cell (or perhaps quantum dot panels). This component would beam energy toward a really huge antenna which would have the function of beaming power back to Earth in the form of RF beams. "Each of those parts can fit on an existing rocket," according to Solaren's CEO Spirnak. "So you don't need to design a brand new rocket…and…you don't need astronauts or robots."[168]

Figure 2. Concept of a Future Solar Power Satellite Configuration in Space (Graphic Courtesy of NASA)

[168] "Company Plans to Pull Solar Energy From Orbit", NPR Story, Dec. 9, 2009, http://www.npr.org/templates/story/story.php?storyId=121531373

Differing Perspectives on Regulating Commercial Solar Power Satellites.

Key Actors	Safety Regs. & Standards	Security Concerns	Interference w/ Other Sat. Applications	Other Regulatory/Trade/ Tax/Environ. Issues
Civil Gov't Agencies	Systems to be un- manned & robotically controlled. Space agencies safety concern will be orbital obstruction & debris. Air traffic regulators will be concerned with radiation damage to passengers	Interference to Deep Space Network, ISS, & government comm. satellite networks. Proximity to manned space stations	The high transmission power levels for SPS could adversely affect sats in lower orbit. Reflected power from ground receivers would be major problem to Comm. Sats. Alternative might be laser transmission.	**Regulation:** Who regulates safety, deployment & oper. **Trade:** Word off world system be subject to WTO-GATS? **Tax:** What type of national & international taxes would apply? **Environment:** Could systems adverse as well as positive environ. Impact? Credit for emission reduction?
Military	Radiation damage to pilots, Interference with vital comm. systems	Interference to military satcoms. Could systems be utilized as a weapons system?	Interference with military space assets	**Regulation:** Who is in charge of regulating and licensing?
Commer-cial	Perspective is that if the systems are robotic there are minimum of safety concerns. Will need to demonstrate space to Earth transmission are non-harmful to passengers, satcom, Deep Space networks, radio astron.	Will need to allay concerns that SPS could be used as a dead ray or weapons system.	Ground array antennas will need to large in area to allow wide beam spreading and designed to not reflect energy into space. Address laser vs RF power transmission.	**Regulation:** Will seek to have minimum of national, regional or international regulation. Need RF allocation for space to Earth transmission. **Trade:** Seek to be free of national & international trade regs. **Tax:** Seek to be free of taxes where ever possible. **Environment:** Seek to have "green" status clearly identified.
International Agencies (ITU,WMO, UNEP, UNO)	ITU, and maybe ICAO, may seek maximum. power transmission levels for this service. WMO & UNEP will examine radiation levels & environmental impacts. COPUOS may explore SPS oper. guidelines	UN Security Council and COPUOS will likely consider possible use of SPS as weapon systems. Consider its consistency with Outer Space Treatry, etc.	ITU will seek recommendation for SPS oper. To minimize interference. May ask for filters, etc.	**Regulation:** ITU, WMO, UNEP, ICAO and UNO will consider to what extent this is a national regulatory vs. a supranational reg. matter. **Trade:** WTO may consider who SPS falls under GATS, **Tax:** National vs. international concerns, if SPS transmits to more than one country. **Environment:** Possible new treaty or guidelines for SPS

Again there may well be divergent opinions on how to best control and regulate this new type of space operation from the commercial ("the merchants"), the military ("the guardians") and the governmental agencies ("the civil space advocates and regulators").

Use of Commercial Systems to Support Military Applications in Space:

Despite the commitment to use space for peaceful purposes under the Outer Space Treaty, the military uses of space have continually expanded over recent decades. We have ICBMs armed with weapons of mass destruction. There are military surveillance satellites that can monitor activities with resolutions measured in centimeters and communications satellites that can provide updated information to war fighters in the field with the latest intelligence acquired from ground sensors, UAVs and even satellites in space. Research has gone forward to explore "star wars" type technology (known formally as the "strategic defense initiative") that includes space-based laser weapons systems, and artificially intelligent space targeting systems. The development of new types of space systems constantly triggers others to duplicate or create even more sophisticated space-based military systems.

As the technology has become more and more sophisticated the tight cooperative relationship between military armed forces and high technology companies has likewise increased. Initially military communications systems, civil governmental communications and commercial satellite systems were quite separate, but in the last decade, in particular, there has been a rapid growth of so-called "dual use" commercial satellite systems. Today commercial satellite systems such as Inmarsat, Intelsat, Eutelsat, SES global, Iridium, Global Star, Thuraya, etc. are carrying up to 80% of military and defense related satellite communications services. The recent launches of the Wideband Global Satellite (WGS) system will lessen this dependence on commercial systems in the short term, but not in the longer term. In the past decade we have seen commercial communications satellite companies become even more directly involved. The XTAR Corporation has deployed capacity in the military X-Band and then sold this to military users. The U.K., France and Italy have now contracted with commercial organizations to develop and deploy specified new military communications satellite systems under contract with excess capacity being made available to support commercial communications requirements.

A host of commercial remote sensing satellites with extremely high resolution (from 80 to 35 centimeters) have now been deployed commercially. These commercial birds such as Geo Eye, Quick Bird, Spot Image, Ikonos, etc. plus many lower resolution systems are being developed and deployed around the world. (See Figure 3) Although there is so-called "shutter control" on many of the commercial systems that can be exercised to prevent imaging in war zones, some of the systems operated from Russia and other countries are not subject to such controls. The bottom line is that more and more commercial systems for satellite communications, remote sensing, and with the launch of the Galileo system, space navigation and satellite targeting systems are being launched into orbit. Commercial space planes, and commercial space stations could also be employed for defense-related or military operations.

Figure 3. The Geo Eye 1 Remote Sensing Satellite Depicted in Earth Orbit (Graphic Courtesy of Geo Eye)

The question is to what extent, over the course of the 21st century, military and defense-related space operations might be "commercialized"? There are a host of questions that follow as to who will control such commercial space military-linked services? Who will provide regulatory oversight at the national and supranational level and what impact will the various major "space actors" make as the process continues in coming decades?

Diverging Perspectives on the Commercialization of Military-Related Space Services

Key Actors	Safety Regs. & Standards	Security Concerns	Impact on Other Sat. Applications	Other Regulatory/Tax /Trade/Environ. Issues
Civil Gov't Agencies	Most "dual use" sat apps. today are unmanned. Related orbital debris is main concern. Commercial space planes or platforms would give rise to a host of new space safety concerns.	There could be demands for more space treaties to limit & control private space planes or platforms.	"Dual use" commercial SatCom, remote sensing, space navigation, and even SPS could make these satellites vulnerable to attack in a time of war.	**Regulation:** May be increased demand for explicit control & licensing for "dual use" space services—particularly for commercial space planes & platforms. **Tax & Trade:** Issues largely resolved except for space planes & platforms. **Environ:** "Dirty space planes" are concern
Military	Safety concerns would focus on private space planes and platforms. Mission critical comm. functions now consigned to some commercial SatCom networks.	Commercial services can add flexibility & cost savings, but lack of command control a concern. Would limit "dual use" if possible.	Military seek priority restoration over other services, but often without key guarantees. Orbital debris endangers both dedicated military SatCom & surveillance as well as critical dual use networks.	**Regulation:** Would like greater control over commercial market and of new space technology through national policy & regs. **Tax & Trade:** Tax & trade concessions can lead to greater control of commercial systems. **Environ:** This is not current military concern
Commercial	Reliability & safety is key to commercial market expansion. Military payment premiums allow for greater safety margins	Commercial suppliers will take lead from military in defining security risks & precautions. Yet would like to retain control over private space planes and platforms.	Governmental and military services represent an expanding market and base for future research to expand regular commercial markets	**Regulation:** Seek minimum regulatory control by civil gov't or military if possible. **Tax & Trade:** Seek to avoid nat'l or internat'l taxes and trade restraints to export technology. **Environ:** Will seek lowest cost & most reliable technology regardless of environ. risk
Internat'l Agencies (COPUOS, Security Council, IAEA,)	Safety, other than orbital debris, not concern. Main objective is to avoid excessive space militarization	Prime concern is global security. Seeks avoidance of space weapons and/or verification of space armaments.	Seek protection of basic space apps. Concern about possible military apps. of SPS & space navig. system	**Regulation:** See ban of all space military systems—gov't or commercial alike. **Tax and Trade:** Use of tax and WTO trade restrictions to limit space militarization. **Environ:** Seek additional controls.

Establishment of Space Colonies and Terra-Forming of Planets

The previous topics considered, such as commercial space transport, new commercial solar power satellite systems, and expanded use of commercial systems to support military applications in space, all present current commercial initiatives. Commercial activity in these areas represents real world markets and current market services or relatively near term programs. The possible creation of space colonies (in orbit or on the Moon) or future prospects of "terraforming" Mars, obviously represent the longer-term future and are much more speculative. Nevertheless, planning exercises are underway. The U.S. space agency NASA has been working with counterpart space agencies since 2004 on implementing a future vision for longer-term initiatives in space. It has, in particular, joined with thirteen other space agencies – the European Space Agency, the French CNES, the Italian Space Agency, the German Space Center DLR, the British National Space Center, the Canadian Space Agency, the Australian Commonwealth Industrial and Scientific Organization, the Japanese Aerospace Exploration Agency, the Chinese National Space Agency, the Korean Aerospace Research Institute, the Indian Space Research Organization, the Russian Space Agency Roscosmos, and the Ukrainian Space Agency - to coordinate exploratory plans and to prepare a "Global Exploration Strategy" that spells out the rationales for carrying out a long-term program of human space exploration.

These efforts to create a future long range exploration plan have more or less assumed the status quo in space with civil space agencies providing the primary leadership in defining objectives and developing the new technology. This, however, may not at all represent the future reality. It is possible that in one scenario the military may play a predominant role in a future where space militarization has progressed well beyond today's limits on weapons systems and military actions in space. Another scenario entirely might be one based on the dynamic growth of space entrepreneurs fueled by the actions of today's innovative "space billionaires" who are seeking to evolve a totally different commercially based space future.

The precedents in space law, treaties and international agreements, space regulations, safety standards and guidelines, tax policies and trade requirements, and even new national and supranational environmental controls will undoubtedly shape future space programs. The extent to which commercial space initiatives succeed in coming decades are likely to be closely related to the progress (or lack of it) achieved in private space travel, solar power satellite systems and use of commercial space facilities to carry out future military space activities. The extent to which commercial organizations play a predominant role (or not) in human space travel (i.e. not only space planes, and private space platforms, but also perhaps in tether lift systems, space elevator systems, etc.) will count a great deal. Likewise the national and international regulatory processes that shape commercial solar-based power and "dual use" military systems will powerfully shape future space policy and regulation.

Actually, to a certain extent there is a four-way struggle to shape the future of space regulation. The commercial sector ("the Merchants"), the military sector ("the Guardians"), the civil government agencies dealing with space ("the National Space Advocates & Regulators") and the International Agencies all have their priorities, their goals and their strengths and weaknesses. The biggest unknown, with the largest upside (or downside) potential is the commercial sector. If the commercial sector can truly generate new and breakthrough technology and capitalize major new initiatives in space, it will change everything—space accomplishments, the speed of progress to create an off-world industry, and the shape and form of space policy and regulation. New types of industrial space consortia based on international partnerships working under international agreements and new forms of space regulation can potentially accomplish a great deal. In the past new types of international space consortium arrangements such as those created with Arianespace, Intelsat, Inmarsat, ILS, Sea Launch and Hispasat/Hisdesat have shown great promise but with varying degrees of long term success.

CONCLUSION

Most attempts to examine international space policy, law and regulation begin by examining the existing space treaties and international agreement. This is a useful way to examine where we have been, but it does not necessarily help to envision the future. An examination of various goals of the major players and an analysis of where these objectives coincide and where they are in conflict is a useful place to start in terms of trying to chart the future of human space activities. Commercial space activities have grown from a modest few millions of dollars in 1965 to a total today of close to a quarter of a trillion dollars (US). With the growth of existing space applications, new space applications such as solar power satellites, the continued expansion of "dual use" commercial systems to support military operations, and the potential growth of commercial space travel, commercial space planes and commercial space platforms offer the prospect of the next fifty years in space being truly spectacular in terms of creating a major "off world" industry. Without the proper development of new space policies, innovative regulatory actions and a creative combination of the various interests of the major "space actors", this could all end with very little being accomplished. It is important to monitor each of these space actors closely. We need to follow closely the announced space policies of the prime space nations, to review the stated strategic goals of the major military interests, to analyze the business plans of the various commercial space interests, and to encourage a creative space policy agenda among the international agencies. Over the last fifty years an amazing amount has been accomplished in space, but the last twenty years have been in many ways a disappointment.

We need creativity of space policy and entrepreneurial initiative. This is essential if we are to see the breakthrough accomplishments to which we aspire. Remarkable things could be achieved in terms of clean and plentiful energy from space, disarmament verification, effective use of space technology to combat "climate catastrophe", and extremely low cost access to space via totally new technology. This might be accomplished through radical innovation such as the so-called "space elevator" or perhaps nuclear propulsion or a rail gun launch system. What seems clear to this author is that in fifty years time the concept that the safest way to send people into space is by putting them on top of a controlled chemical explosion will seem quaint indeed.

There is clearly much more that needs to be done to examine the goals and objectives as well as the programs and performance of the key "space actors". The above charting process is only a rudimentary methodology, but it does represent a start to a clearer insight as to the development of future space policy, regulation and the course of space industrialization.

Acknowledgments

I would like to thank Dr. Ram Jakhu of McGill University and Peter Marshall of Dorset, United Kingdom for their review and comments to help improve the accuracy and currency of information contained in this article.

In: Space Policy and Its Ramifications
Editor: John P. Ramos

ISBN: 978-1-61761-555-9

Chapter 5

SPACE POLICY FOR LATE COMER COUNTRIES: CASE OF SPACE INDUSTRY CLUSTER IN SOUTH KOREA

Joosung J. Lee and Jeongmook Kim
Korea Advanced Institute of Science and Technology,
Daejeon, Republic of Korea
Yonsei University, Seoul, Republic of Korea

ABSTRACT

Korea's space development program was created almost 40 years behind advanced countries Nevertheless; it has been making remarkable growth including the construction of Korea 'NARO' space center and the possession of 10 satellites developed by Korea's own technology.

There are; however, a number of problems in Korean space development policy—dependence on imported core technology, uncertainty of budgetary allocations that are influenced by political or economical circumstances, insufficiency of human resources, concentrated authority on KARI(Korea Aerospace Research Institute), and weakness of aerospace industry infrastructures. They all must be re-aligned in order to develop long-term space industry strategies.

In this study, we analyzed the current space policy issues mentioned above by comparing Korea and other countries' space industries and government policies. In addition, we have suggested that the private sector should vitalize the space industry in terms of establishing a functional space industrial cluster. That is, Korea should establish a functional space industry cluster and remove inefficient, overlapping investments.

For successful space development, late comer countries should institute a privately financed space development policy. The establishment of a space industrial cluster policy in this study is expected to help invigorate the space industry of Korea and catch-up countries with a similar environment.

1. INTRODUCTION

1.1. Overview and Structure of Research

In the Republic of Korea, working processes with regards to the aerospace industry, such as space development planning, policy implementation, and R&D to launching of the spaceship, have been led by the government. This planned aerospace industry development by the government can be effective for a short term; however, it may lack consistency in financial and political support when it faces political and economic influences and when the government does not have a clear goal. Therefore, in order to solve this problem, private sectors or enterprises should lead the aerospace industry, and eventually, this method should become effective for a long term. This article suggests a cluster-structure method by examining each function of the aerospace industry, which favors a privately-led space development policy.

1.2. Definition of the Aerospace Industry

Most of the academic literature and law related to the aerospace industry explicitly refers to air and space all together since these two terms have a strong relation to each other. Article 2 of the Korean Aerospace Development Act (amended on Feb. 29, 2008) prescribes that aerospace industry refers to the industry related to producing airplanes, aerospace ships, and relevant avionic devices or materials, and it also refers to the applied technology industry of airplanes and aerospace ships defined under the Decree of the Ministry of Knowledge Economy. This article will discuss only the aerospace industry, excluding the airplane industry. Therefore, in this article, the aerospace industry means the industry with regards to producing aerospace ships and related avionics and materials and the applied technology industry of aerospace ships. In addition, the scope of the industry in this article includes the relevant business for aerospace development in advance (pre-development) and applied technology business after development (post-development).

Aerospace can be categorized as the aerospace manufacture industry, aerospace service industry, and aerospace applied industry, and each sector can be again subcategorized by its unique technological feature. Satellite stations, projectiles, and ground facility manufactures is classified as the aerospace manufacture industry. Communication and broadcasting, observation and control, and aerospace transportation is categorized as the aerospace service industry. Projectile support, material analysis, ground experiments, and aerospace insurance is categorized as the aerospace applied industry. This classification method can be summarized as shown in Figure 1-1.

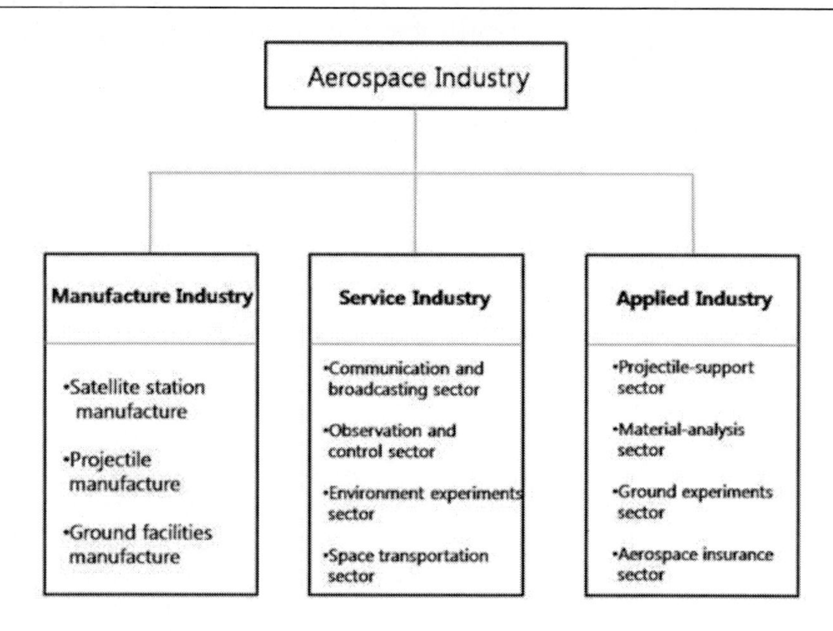

Figure 1-1. Categories of the Aerospace Industry

1.3. Importance of Aerospace Industry

1.3.1. Significance of aerospace industry

A high value technology-focused industry is very essential to a country, such as Korea, which does not have sufficient natural resources. Aerospace industry development can contribute to the national economy by the growth of machinery, electric, and electronic industries. In addition, it can activate not only research activities at universities and research institutions, but also improve the infrastructure of aerospace development. The importance of the aerospace industry can be analyzed by two major subjects, the influence on the state economy and the influence on technology.

1.3.1.1. Influence on national economy

One of the most important influences of the aerospace industry is its effects on the state economy. The aerospace industry is the assembled business of technology-concentrated industries that needs applications of machinery, electronic, and materials of high technology, and it is a high value industry that may create beneficial economic outcomes in applied technology industries after its development. Considering this with regards to the Naro aerospace ship case, its economic benefit from the Naro project is estimated at about 1.8 trillion to 2.4 trillion Korean won (Korea Industry Institute, 2009) which can be compared to the economic benefit of the World Cup or the Olympics. Moreover, it can create employment for 7,689 people, including 4,600 who are related to the R&D of projectiles. When estimating and considering this positive effect in one sector, projectile, it is very certain that the overall aerospace industry can create significant economic benefits.

1.3.1.2. Influence on technology

Applied technologies in the aerospace industry are very useful in commerce. In fact, the influence of the aerospace industry on the economy is almost three times as much as that of the automobile industry. For example, not only are medical devices, such as the eczema laser at ophthalmologists and dermatologists, the CT scanner, and the MRI commercialized-technology products, but also retorts used for food products. Developed materials, technologies regarding machinery, system development technologies, and manufacture technologies in the aerospace industry can create more influence on the technology foundation sector and applied technology sector (see Table 1-1 below).

Table 1-1. Effectiveness of the Aerospace Industry

Category	Aerospace Technology	Other Applied Technology Sectors
Material Devices Miniaturization, Microlight, Environment energy sensor	Direct circuit, Honeycomb materials, Insulator, Heat pipe, Composite, Solar battery, Fuel cell, Rocket engine, Diverse position controls, Environment center	Direct circuit, Micro medical device, Avionics, Housing insulator, High efficient heat exchanger, Airplane frame, Device materials, Fuel cell, Solar battery, Automobile collision sensor, Environment sensor
System Technology Control technology, Communication technology, Image processing technology, Computer technology	Inertial navigation system, Route guidance technology, TM/TC data transport, Digital image processing, Miniaturized computer manufacturing	Inertial navigation system for private airlines, Remote-emergency medical data communication, High density transport system, Chromosome disorder detector, Bank and large business information system
Production and Manufacture Design technology, Manufacturing and assembling, Experiments inspection	Structure Analysis, Simulation, Clean room measurement, Environment experiment, Defects sensor, Planning and scheduling process	Civil engineering, Architecture simulation, Virtual experience, Medical and semiconductor clean room, Industry measurement, Medical ultrasonic analysis, Engineering Management, Planning and scheduling process

Sources: NASA, Spinoff (1979), and Japan Aerospace Industry Association, Japanese Aerospace Industry (1984)

2. Current Issues in the Aerospace Industry and Policy Analysis

2.1. The Aerospace Industry in Korea

2.1.1. Scale of the aerospace industry in Korea

When dividing the Korean aerospace industry into two periods, beginning (period 1) and developing (period 2), the U.S., Japan, and the EU already established a structure for the aerospace industry in the 1990s through their government-led policies in the period 1. Moreover, these regimes expect that their aerospace industries will be commercialized by about 2010 (see Figure 2-1). On the contrary, Korea started their aerospace industry, which was led by the government, around 1990, but its stage still remains as beginning and immature. This implies that because Korea has a 40-year gap in development from the U.S. and Japan, it will not be easy to overcome the gap, unless the Korean government supports the aerospace industry with continuous investment, innovative strategies, and active policy support. When looking at the turnover of each aerospace industry sector in 2007, it made profits of 456.2 billion Korean won in the aerospace device manufacturing sector and of 806.8 billion Korean won in the applied aerospace technology sector, whose total profit was 1.263 trillion Korean won. The percentages of each sector are 34% and 64%, respectively (Ministry of Knowledge Economy, 2008, Research of Aerospace Industry, 2009). This outcome indicates that Korea focuses on profits from the applied technology industry sector through using existing aerospace devices rather than profits through obtaining core-technology for innovating aerospace devices. The satellite broadcasting sector also created the largest amount of profits, 40% in overall aerospace industry; the following sectors, which made profits, are satellite communication (19.2%) and satellite (19.1%). The unique feature of profits from certain sectors, such as satellite broadcasting, satellite communication, and satellite, shows that Korea has developed certain sectors focused on short-term profits and public demands.

The import and export of each sector of the aerospace industry is shown in Table 2-1, below. The satellite communication sector has made the largest export in the Korean aerospace industry, which was 75% of the total exports; the profit is approximately 43 billion Korean won. On the contrary, the largest import sector is the satellite and ground facility sector, whose percentage of the total import is 88%, approximately 110 billion Korean won. In short, the total profit from export is 57 billion Korean won, but the amount of import is 124 billion Korean won, which means the current aerospace industry makes a loss. Among the overall aerospace sectors, the satellite sector indicates the third largest in export but the largest in import. This proves that the satellite sector in Korea heavily relies on foreign technology in core technologies. Currently, the Ministry of Knowledge Technology announced that it will entrust domestic private enterprises with the task of manufacturing multi-purpose satellite, 3A's frame. This means that private enterprises will take part in satellite technology development, thereby exporting technologies and Korea's independence in the aerospace technology industry.

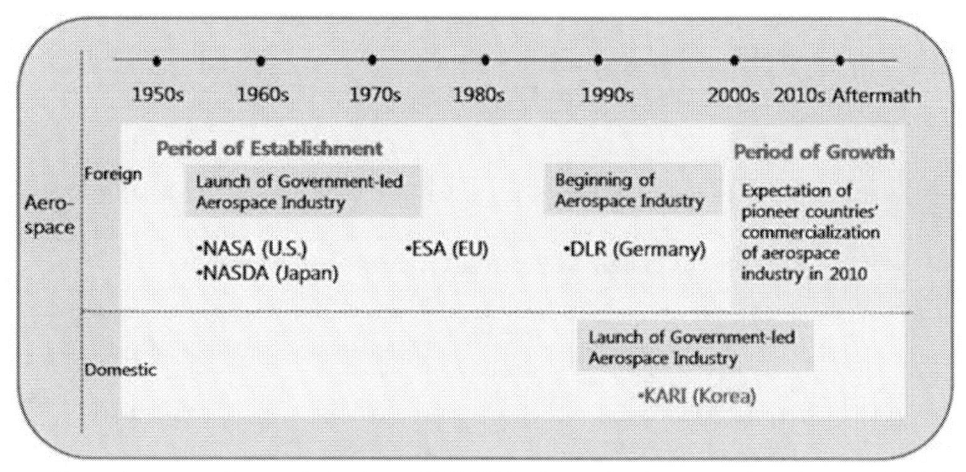

Source: Korea Aerospace Research Institute, 2006 White Paper for Aerospace Development (2007)

Figure 2-1. Stages of Aerospace Industry

Table 2-1. Export and Import in the Aerospace Industry Sectors (million Korean won)

Category	Subcategory	Export		Import	
		Won	Percentage	Won	Percentage
Aerospace Device Manufacture	Satellite	10,697	18.7%	64,630	52.1%
	Projectile	0	0.0%	7,994	6.4%
	Ground Facility	3,205	5.6%	44,875	36.2%
Total		13,902	24.2%	117,499	94.8%
Applied Technology Aerospace Industry	Remote-Control	345	0.6%	674	0.5%
	Satellite Navigation	0	0.0%	348	0.3%
	Satellite Broadcasting	0	0%	0	0%
	Satellite Communication	43,101	75.2%	5,273	4.3%
	Space Science	0	0.0%	169	0.1%
	Other	0	0.0%	0	0.0%
Subtotal		**43,446**	**75.8%**	**6,464**	**5.2%**
Total		**57,348**	**100.0%**	**123,963**	**100.0%**

Source: Ministry of Knowledge Technology(Korea), 2008 Research of Aerospace Industry (2009)

2.1.2. Issues of aerospace enterprises in korea

The scope of capital investment of 34 Korean enterprises whose sales related to the aerospace industry and employment is shown in Table 2-2, and the enterprises' turnovers are indicated in Table 2-3.

Table 2-2. Scope of Fund of Aerospace Enterprises

	Less than 1 billion Korean won	1 billion to 10 billion Korean won	10 billion to 100 billion Korean won	100 billion to 500 billion Korean won	Over 500 billion Korean won	N/A
Number of Firms	9	11	1	8	2	3

Source: Ministry of Knowledge Technology(Korea), 2008 Research of Aerospace Industry (2009)

Table 2-3. Scope of Profits of Aerospace Enterprises

	Less than 1 billion Korean won	1 billion to 10 billion Korean won	10 billion to 100 billion Korean won	100 billion to 500 billion Korean won	500 billion to 1 trillion Korean won	Over 1 trillion Korean won
Number of Firms	3	9	8	5	3	4

Source: Ministry of Knowledge Technology(Korea), 2008 Research of Aerospace Industry (2009)

The total number of enterprises that have funds below 10 billion Korean won was 24 (67.6%), and the percentage of firms with over 10 billion Korean won was 58.8%. This indicates that the majority of aerospace enterprises in Korea are small and medium-sized enterprises, rather than conglomerates. The Korean government has restricted large firms' participation in the aerospace industry by means of government-led policies and laws, which reduced these firms' incentive for entering the aerospace business; thus, firms in the military and defense industries were more likely to enter the aerospace industry. Accordingly, it is possible that the system change from government-led to private cannot easily solve the problem of large firms' refusal to enter this business. This means Korea may have problems of obtaining large assets and funds for aerospace industry development.

2.2. Global Investments in the Aerospace Industry

The total amount of estimated global investments in the aerospace industry in 2008 was 62.1 billion USD, which can be divided into 32.7 billion USD on the private sector and 29.4 billion USD on the military aerospace programs. In particular, the U.S. invested 18.6 billion USD on a private sector, which was 56% of the total budget in sum of other governments. Except for the U.S., the countries of Japan, France, Russia, Germany, China, and Italy invested more than 1 billion USD by private sector, and India also invested 0.96 billion USD by it. The percentage of investments by the ten countries, which invested the most on the private aerospace sector, is 90% and the other 30 countries invested a total of 2.7 billion USD.

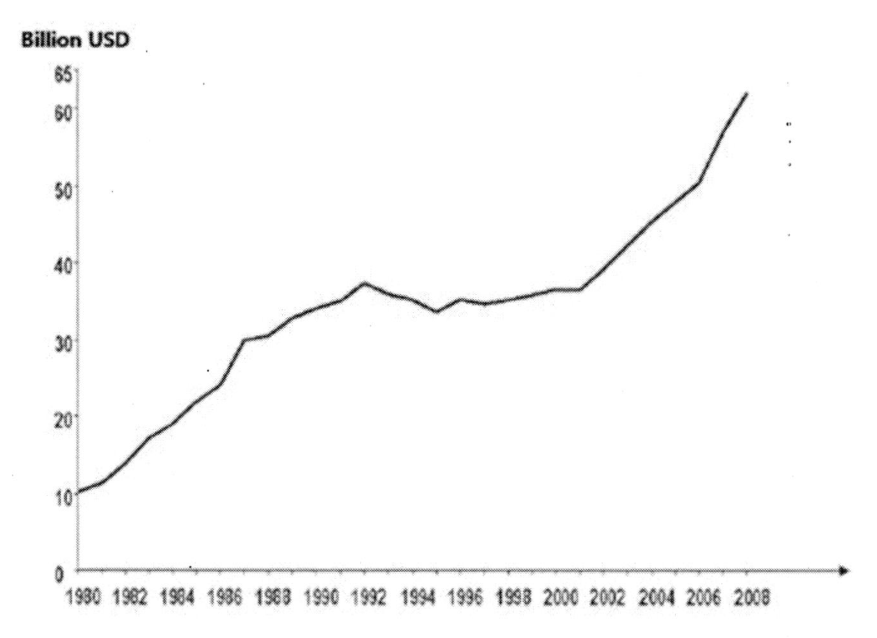

Source: Euroconsult (2008)

Figure 2-2. Governments' Investments on the Aerospace Sector

Table 2-4. Major Countries' Budgets for Aerospace Industry (billion USD)

	2003	2004	2005	2006	2007	2008
U.S.	32,886	34,622	36,693	39,082	41,996	46,385
Japan	2,834	3,102	2,933	2,744	2,631	2,948
France	1,967	2,183	2,252	2,290	2,474	2,694
Russia	350	556	741	1,050	1,280	1,465
Germany	914	1,031	1,004	1,061	1,299	1,452
China	913	960	1,082	1,161	1,231	1,300
Italy	802	811	888	830	1,064	1,232
India	489	561	610	662	926	966
U.K.	374	677	682	693	763	759
Canada	201	220	238	277	338	358
Korea					317	287

Source: Euroconsult (2008)

Table 2-5. Profits in the Aerospace Industry by Sector

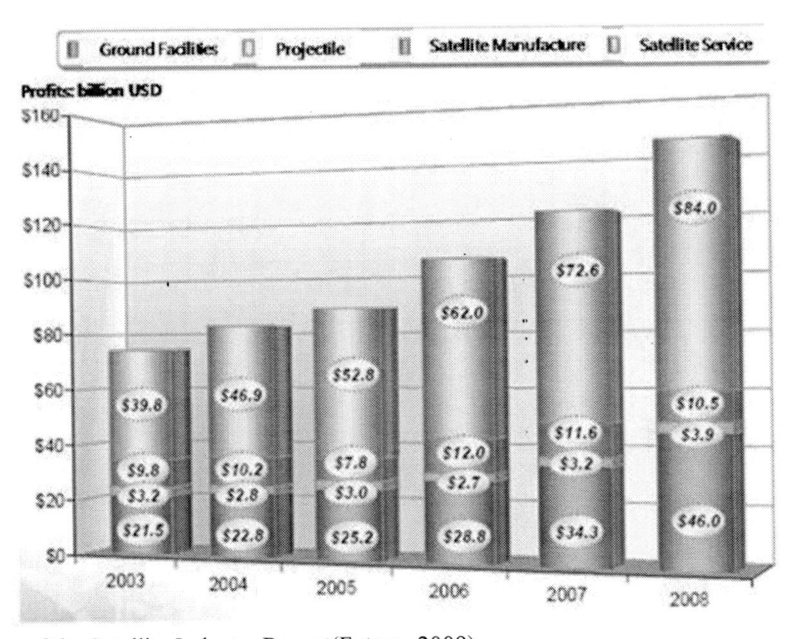

Source: States of the Satellite Industry Report(Futron, 2009)

In the order of investment amounts on the aerospace industry in both private and military sectors, there are ten countries, which are the U.S., Japan, France, Russia, Germany, China, Italy, India, the U.K., and Canada. In particular, the U.S., Russia, Japan, and China showed their increase of investments in 2008 by 10%, 14%, 12%, and 6%, respectively. The amount of the U.S.'s investment in aerospace was 75% of the total investments of all other countries. By categorizing based on aerospace institutions, the U.S. Pentagon, U.S. NASA, EU ESA, Japan JAXA, France CNES, Russia Roscosmos, and China CNSA invested more than 1 billion USD in 2008 (see Table 2-4). In 2007, the total amount of sales in the aerospace industry worldwide was 123 billion USD. The satellite service sector, among others, such as ground facility, projectile, satellite manufacture, and satellite service, made the highest profit, which was 60% of the total amount, and it increased 18% compared to the previous fiscal year.

Examining the turnover of each aerospace industry sector in Korea, the satellite broadcasting sector made a profit of 40% of the total, and this indicates that most of the profits were earned in the applied aerospace technology sector rather than aerospace device manufacturing. According to Table 2-5, the percentage of global aerospace profit in the satellite service was about 60%. Therefore, it is crucial to not only develop technology in aerospace device manufacturing for the applied aerospace technology sector as a core assignment in this industry, but also to design diverse applied aerospace service models for a long-term commercial profit perspective.

2.3. Problems in the Korean Aerospace Policy and Business Performance

Until now, most of the aerospace industry sectors in Korea were led by the government. In particular, when private enterprises took part in aerospace projects, they sometimes had the burden of a certain percentage of investment costs. This system made these firms face difficulties in their budgets, which eventually hindered their interests in investments. For example, it is almost impossible to recoup firms' investments on multi-purpose satellites 1 and 2 projects because of the accumulated large burdens of investment costs. In addition, the discontinuity in aerospace development makes it more difficult to manage human resources and organizations. Moreover, it is very difficult for a private enterprise to use expensive experiment devices and materials that belong to the government. Therefore, the government needs to establish a program of usage fee standards and device/material usages, in order to support private enterprises. Small and medium-sized enterprises which entered the aerospace industry later have more difficulties since the government's current development system is focused more on incumbent enterprises. Furthermore, the government sometimes disregards and excludes business-conduct experiences and built-up technologies in its succeeding or relevant aerospace projects for a government policy reason; thus, these accumulated experiences and technologies seem to vanish. The Korean government faces difficulty in attaining core technology since developed countries are reluctant to transfer their technologies. In addition, most of the advanced technologies were obtained by scientists and engineers supported by the government, and this system blocks technology transfer from the government to the private sector. Moreover, it is difficult to observe synergy effects from nation-wide aerospace industry development because of the dearth of information and technology exchanges between institutions and businesses in Korea.

The Korean government set up two major departments regarding aerospace industry, the "macro-science policy department" and the "aerospace development department" at the macro-science support division in the Ministry of Education, Science, and Technology. These two departments are responsible for establishing and implementing the government's aerospace policy. In addition, the Korea Aerospace Research Institute (KARI) is in charge of aerospace industry development, including satellite manufacturing and astronaut training. However, the Korean administrative bodies for aerospace development have the problems of insufficient support. Moreover, KARI is deficient in human resources, budget and administration systems, and researchers, as compared to other developed countries' aerospace policies and the size of their relevant departments. This makes it more difficult for Korea to be one of the top ten countries in the aerospace industry in 2025. The moving of government officers to another department also creates a lack of specialty in aerospace developments. Therefore, the government needs to establish a specialized department or institution for ensuring future planning and continuous policy processes.

In demand perspectives, the Korean government should implement stable and continuous national business plans in order to create a foundation for the aerospace industry development. It is very crucial to guarantee steady demands for the aerospace industry because of its unique requirement of a large capital investment on projects. Otherwise, the aerospace industry will not be further improved. Therefore, the government needs to offer policy supports and create demands for aerospace industry development.

3. PROPOSALS FOR ENCOURAGING PRIVATE SECTORS' PARTICIPATION IN THE AEROSPACE INDUSTRY

3.1. The Need for Private Enterprises' Participation

According to the Futron Consulting Corporation's 2008 reports about the world competitiveness index in the aerospace industry, advanced countries in the aerospace industry, such as the USA and Russia, demonstrated their high competitiveness in both government and industry sectors (see Figure 3-1 below).

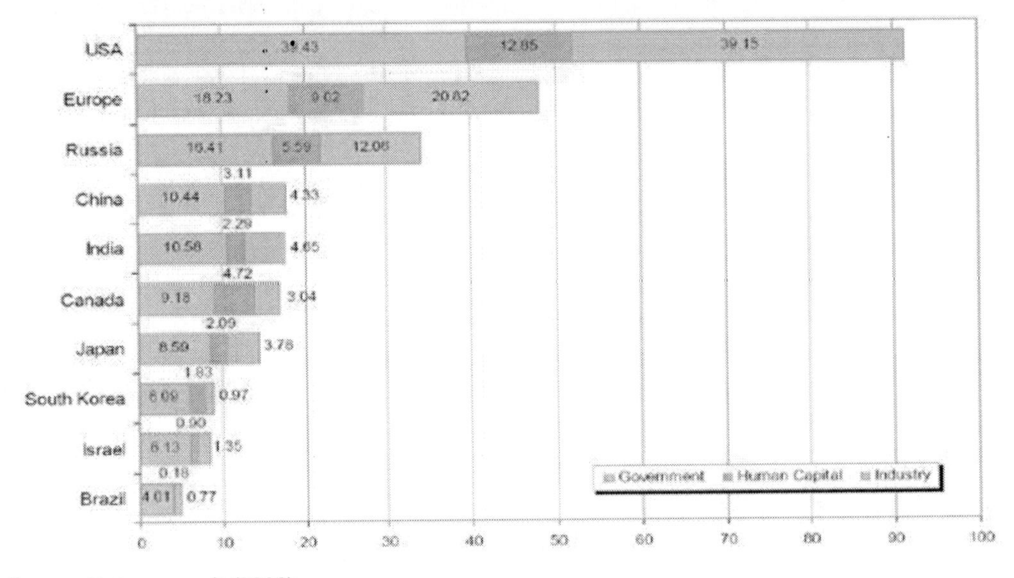

Source: Futronconsult (2008)

Figure 3-1. Space Competitiveness index: Total Aggregate Scores by Country

The index indicates that both public (government) and private (industry) sectors in the USA have almost equivalent strength in their competitiveness, as shown in the index numbers above, 39.43 and 39.15, respectively. However, the aerospace industry in Europe and Russia shows significant strength in the private, rather than the public, sector.

This proves that, in these regimes, competent private enterprises, which have sufficient technology and capital for aerospace development, vigorously take part in the aerospace industry. However, in the case of Korea, the significant gap between public and private competitiveness, as indicated in their 6.09 and 0.97 index numbers, respectively, shows an overwhelming government-led aerospace policy.

In sum, it is apparent that the Korean government needs to aggressively participate in aerospace development; however, private enterprises' involvement and the government's supports for them are also essential in order to improve the competitiveness of the Korean aerospace industry.

3.2. Proposals of Cluster-Structure Policy for Encouraging Private Sectors' Participation

3.2.1. Background for cluster-structure policy proposals

This article suggests a cluster-structure method in order to encourage private sectors' participation in the aerospace industry. In fact, quite a number of regional government bodies actively try to establish aerospace industry areas and help aerospace-relevant enterprises join the regional governments' projects by moving in their regions. If the central government does not control these diverse projects by its policy, it is possible that each regional government's effort will not benefit Korea as a whole.

The cluster policy does not suggest a new policy instrument but proposes the regional integration of particularly related industries for core technology development, which is essential for industrial competitiveness improvement, thereby creating various values and networks and also offering competitive and cooperative environments and systems in these regions.

Most of the clusters, which have been currently created in Europe, retain characteristics of a science-based cluster, and they rely on new technologies and knowledge and also result in spin-offs of universities and research institutes. The case of Europe can be a good example to Korea, since most of the participating firms in the Korean aerospace industry are small and medium-sized enterprises, and the European method is suitable for the current Korean system.

3.2.2. Conception of cluster

The word "cluster" has been deeply discussed since Porter (1990) emphasized the role of industrial clusters for improving competitiveness, as illustrated in his book "The Competitive Advantage of Nations." He explains that cluster is the spatial integration of competitively and cooperatively correlated firms, specialized suppliers, service suppliers, relevant industrial entities, and relevant organizations. However, the concept of cluster is not a simple "integration" but the objective for enhancing synergy effects through each entity's regional integration, similarity of organization, and network for substitutability.

Furthermore, Enright (1996) illustrates that industrial cluster is the combination of each enterprise and supporting organization, which is an important factor for each enterprise's competitiveness. He explains the elements for merging them to the cluster are purchaser-supplier relationship, joint technology, mutual demanders and distribution channels, and a common labor pool.

As cluster becomes one of the most important factors for a state or city's competitiveness, the paradigm of the industrial economic policy changes so that it is cluster-based. Since knowledge-based economy has become one of the crucial factors, competitiveness of the economy and industry become heavily influenced by technological development and innovative capacity.

Moreover, industrial policy, when cluster-focused, has a distinctive feature of systemic advantage, which integrates certain regions' production, research, and supporting function for maximizing competitiveness, thereby exploiting the external economy. In the case of the USA and the EU, their regimes have improved their regional and state economy development and enhanced their competitiveness in the aerospace industry through establishing 40 clusters in the USA and Regional Innovation System(RIS) and Regional Innovation and Technology Transfer Strategies and Infrastructures(RITTS) in the EU.

3.2.3. The effects of cluster

a. Innovation Through Cluster

Cluster can increase innovation, enhance knowledge extension and knowledge accumulation, and also improve a rapid knowledge-ripple effect between enterprises. Since the firms within a cluster reside next to each other, they can very easily cooperate with each other for innovation and can discuss changes in characteristics of inputs, as well as exchange information about new products and perceived changes in market demand.

Notteboom (1998) argues that geographic proximity is a crucial element for rapid innovation changes, and Baptista and Swan (1998) also analyze the relationship between enterprise-level cluster and innovation. Accordingly, innovation, knowledge, and technology-ripple effect are the most important factors for innovation development by firms in a cluster.

b. Creation of Competitive Advantages Through Cluster

Porter (1990, 1998) asserts that cluster can strengthen the competitiveness of domestic enterprises and industries in the global market. In other words, firms in a cluster can strengthen their status in the market through scale and scope of economies, thereby reducing costs. Furthermore, the competition between firms in a cluster can enhance efficient production, and this can create their comparative advantages in the global market.

c. Other Effects

Swann (1988) illustrates advantages and disadvantages of a cluster from the enterprises' point of view with regards to supply and demand (see Table 3-1 below).

Table 3-1. Advantages and Disadvantages of Cluster

	Supply Side	Demand Side
Advantages	• Accessibility to Customers • Increasing Market Shares • ost Reductions of Searching Customers • Externalities of Information	• Knowledge-Ripple Effects • btaining Specialized Trainees and Professionals • Acquiring Infrastructure • Externalities of Information
Disadvantages	• Vigorous and Excessive Competition Between Enterprises in the Production Market and Decrease of Profits	• Increases of Input Costs (Land and Labor) from Competition

Source: Swan (1998)

4. A Case Study: Aerospace Cluster in Bayern State, Germany

4.1. Overview of Aerospace Cluster in Bayern, Germany

Bayern State in Germany is one of the most specialized and high-technology aerospace clusters not only in the private sector, but also in the research sector. Bayern is well equipped with aerospace system enterprises, technology facility enterprises, technology service enterprises, research and education institutes, and air traffic facilities, whose structure is ideal for the aerospace industry. The areas nearby Bayern are also exceptionally developed in metal, electric and electronic, and mechanical engineering, which are essential for the aerospace industry.

The major sectors in this area are the production of military airplanes, avionics, and helicopters and also of sub-systems and relevant components. In addition, structures for aerospace ships and satellites, motors, and solar batteries are the important sectors in the Bayern aerospace industry. The current employees of the aerospace industry in Bayern are 23,000 professionals, one third of the total aerospace workers in Germany, and the yearly turnover in this area is approximately 4.8 billion EUR.

4.2. Analysis of the Bayern Aerospace Industry

a. Aviation and Air Industry

One of the headquarters of the Airbus mother company, European Aeronautic Defense & Space Company (EADS), the major competitor of Boeing, is in Munich; and the headquarters of large firms, such as Eurocopter and MTU Aero Engines, and other competent small and medium-sized enterprises, such as IABG, Liebherr Aerospace, and Diehl, are also in Bayern.

Bayern is a leading state in military airplane production. The final assembled works of Eurofighters are made in Manching, and most of the important components of airplanes are also produced in Augsburg. Not only NH90 and Tiger, but also private helicopters, are manufactured in this area. In particular, through the cooperation between Germany and Russia, MIG 29s are maintained and repaired in Manching. Bayern State also holds a key post in Airbus production. Most of the components of Airbus airplanes are manufactured in Bayern, and enterprises in this area produce landing gears, turbines, and cabin systems.

b. Space Industry

Major space enterprises, in particular, EADS-Astrium, MAN-Technologie, and Kayser-Threde, have their headquarters in Bayern. Main components for the Europe rocket boosters, Ariane, such as solid fuel boosters and liquid fuel boosters, are also produced in Bayern. This state is competent and has high technology in ceramic materials and solar panel sectors.

The German Aerospace Center (DLR) in Bayern was authorized to work as a central controller of the Columbus lab, the international space station (ISS), which is the so-called "lab of Europe," in March 2003. The German Space Operations Center (GSOC) is also in Bayern.

c. Research Institutes of the Aerospace Sector

Widely spread research institutions also offer outstanding conditions for aerospace studies. The most important institution and lab are the German Aerospace Center and Brown Hoffer Institute, respectively. In addition, a number of scholars research at the space physics institute, the Max Flank Institute, and the European Southern Observatory. Research on Micro and Nano satellites, which has received attention for its potential in aerospace technology, is also conducted in this state.

Since 1990, the Bayern State government has invested 180 million EUR on aerospace research. Most of the investment was used for cooperative research projects, thereby strengthening cooperation between aerospace enterprises. Educational institutions, such as the Munich Institute of Technology, Erlangen University, and the Federal Military University, have continuously produced high-level professionals, and even Munich only produces approximately 200 students who specialize in an aerospace major.

5. Invigorating Cluster Policy and Its Expected Effects

The governmental role, suggested for invigorating cluster, is following, and the Korean government needs to provide detailed supporting policies for this.

- Intermediary supports of technology and labor for cluster firms' technological development
- Marketing for encouraging foreign direct investments on clusters
- Support for cooperative bodies that can combine cluster participants
- Its role as a purchaser of products and services of cluster
- Standardizing and certifying production of products by cluster
- Supports for participating enterprises by a cooperation with regional governments (tax reduction, investment supports, establishing infrastructure)

We can expect the following effects through cluster in the aerospace industry.

- Creating policies for efficient and effective aerospace development
- Effective and fast supports on participating enterprises through cluster meetings
- Increase of productivity through complementary activities of other cluster participants because each cluster has enterprises that have similar specialism
- Improving competitiveness of major strategic industry, heightening industrial structure, and contributing to employment and regional economy development through creating an infrastructure for the aerospace industry, such as motivating relevant enterprises, research institutions, and integration of professionals
- Establishing a cluster structure for vitalizing the industry is future-oriented because a long-term aerospace development should be private-led with government-support methods

CONCLUSIONS

Korea's space development program was created 40 years behind advanced countries Nevertheless; it has been making remarkable growth including the construction of Korea 'NARO' space center and the possession of 10 satellites developed by Korea's own technology.

There are; however, a number of problems in Korean space development policy - dependence on imported core technology, uncertainty of budgetary allocations that are influenced by political or economical circumstances, insufficiency of human resources, concentrated authority on KARI(Korea Aerospace Research Institute), and weakness of aerospace industry infrastructures. They all must be changed in developing long-term strategies for space policy.

In this study, we analyzed the current space policy issues mentioned above, by comparing Korea and other countries' space industries and government policies. In addition, we have suggested that the private sector should vitalize the space industry in terms of establishing a functional space industrial cluster. That is, Korea should establish a functional space industry cluster and remove inefficient, overlapping investments.

For successful space development, late comer countries should institute a privately financed space development policy. The establishment of a space industrial cluster policy in this study is expected to help invigorate the space industry of Korea and catch-up countries with a similar environment.

REFERENCES

Baptista, R. & Swann, P.(1998). "A Comparison of Clustering Dynamics in the US and UK Computer Industries", *Journal of Evolutionary Economics, Vol. 9.*

Choe, Nammi(2009). "*Current statuses of world governments' expenditures for space program and space markets*", The aerospace industry technology trends ,vol. 7 ,No.1. pp.3-10.(KARI)

Choe, Sumi(2007). "*Development of international space research, and industry trends*", The aerospace industry technology trends ,vol. 5 ,No.1. pp.3-10.(KARI)

Dong-Kyu, An et al.(2007). "*Innovation cluster & regional development*", Sowha.

Euroconsult, (2008).*World Prospects for Government Space Markets.*

Futron, (2008). "*State of the satellite industry report*".

Futron, (2009). "*State of the satellite industry report*".

Japan Aerospace Industrial Association(1984). "*Japan's aerospace industry* ".

Kodai Domihoomi et al.(2009). "*Japan vs. China, space development*", Korea, nBook.

Korea Aerospace Research Institute(2008). "*A plan for establishment aerospace development innovation system & enhancing efficiency international cooperation*"

Korea Institute for Industrial Economics & Trade(2009). "*Economic effects & development task in Launching Naro-1*".

Korea Ministry of Education, Science & Technology (2007). "*Feasibility study for establishment of master plan development promotion*".

Korea Ministry of Education, Science & Technology (2008). "*Research on the actual condition for space industry*".

Korea Ministry of Education, Science & Technology (2009). "*Implementation Plan for Space development in 2009*".

Korea Ministry of Education, Science & Technology(2006). "*Space development white paper*".

Korea National Science &Technology Council(2008). "*The direction of allocation for National R&D budget 2009*".

NASA(1979). "*NASA spinoff 1979*", annual report.

Porter, M. E. (1990). "*The competitive advantage of nation*", NewYork, The Free Press.

Poter, M. E. (1998). "*Clusters and the new economics of Competition*", Harvard Business Review, Nev-Dec 1998.

Science & Technology Policy Institute, Korea(2006). "*Study on the efficient implementation mechmaism for Korea's development*".

Seong Whan Ju(1995). "*Space development and the future world*", Jinhan publisher.

South Gyeongsang Province(2008). "*3rd revision for comprehensive plan at South Gyeongsang Province (2008~2020)*", <"http://www.gsnd.net/99_etc/total_plan/4_1.pdf">.

Swann, G. M. (1998). *Towards a Model of Clustering in High-Technology Industries*, in the Dynamics of Industrial Clustering, Oxford University Press.

In: Space Policy and Its Ramifications
Editor: John P. Ramos
ISBN: 978-1-61761-555-9

Chapter 6

U.S. CIVILIAN SPACE POLICY PRIORITIES: REFLECTIONS 50 YEARS AFTER SPUTNIK*

Deborah D. Stine

ABSTRACT

The "space age" began on October 4, 1957, when the Soviet Union (USSR) launched Sputnik, the world's first artificial satellite. Some U.S. policymakers, concerned about the USSR's ability to launch a satellite, thought Sputnik might be an indication that the United States was trailing behind the USSR in science and technology. The Cold War also led some U.S. policymakers to perceive the Sputnik launch as a possible precursor to nuclear attack. In response to this "Sputnik moment," the U.S. government undertook several policy actions, including the establishment of the National Aeronautics and Space Administration (NASA) and the Defense Advanced Research Projects Agency (DARPA), enhancement of research funding, and reformation of science, technology, engineering and mathematics (STEM) education policy.

Following the "Sputnik moment," a set of fundamental factors gave "importance, urgency, and inevitability to the advancement of space technology," according to an Eisenhower presidential committee. These four factors include the compelling need to explore and discover; national defense; prestige and confidence in the U.S. scientific, technological, industrial, and military systems; and scientific observation and experimentation to add to our knowledge and understanding of the Earth, solar system, and universe. They are still part of current policy discussions and influence the nation's civilian space policy priorities — both in terms of what actions NASA is authorized to undertake and the appropriations each activity within NASA receives. NASA has active programs that address all four factors, but many believe that it is being asked to accomplish too much for the available resources.

Further, the United States faces a far different world today. No Sputnik moment, Cold War, or space race exists to help policymakers clarify the goals of the nation's civilian space program. The Hubble telescope, Challenger and Columbia space shuttle

* A version of this chapter was also published in *Space Policy and Exploration*, edited by William N. Callmers published by Nova Science Publishers, Inc. It was submitted for appropriate modifications in an effort to encourage wider dissemination of research.

disasters, and Mars exploration rovers frame the experience of current generations, in contrast to the Sputnik launch and the U.S. Moon landings that form the experience of older generations. As a result, some experts have called for new 21st century space policy objectives and priorities to replace those developed 50 years ago.

The authorization of NASA funding in the National Aeronautics and Space Act of 2005 (P.L. 109-55) extends through FY2008. Congress may decide to maintain or shift NASA's priorities during the next reauthorization. For example, if Congress believes that national prestige should be the highest priority, they may choose to emphasize NASA's human exploration activities, such as establishing a Moon base and landing a human on Mars. If they consider scientific knowledge the highest priority, unmanned missions and other science-related activities may be Congress' major goal for NASA. If international relations are a high priority, Congress might encourage other nations to become equal partners in NASA's activities. If spinoff effects, such as the creation of new jobs and markets and its effect on STEM education are Congress' priorities, then technological development, linking to the needs of business and industry, and education may become NASA's primary goals.

INTRODUCTION

Current U.S. space policy is based on a set of fundamental factors which, according to an Eisenhower presidential committee, "give importance, urgency, and inevitability to the advancement of space technology."[1] These factors were developed fifty years ago as a direct result of the Soviet Union's (USSR) launch of the first artificial satellite, Sputnik. This launch began the "space age" and a "space race" between the United States and USSR.

The four factors are the compelling need to explore and discover; national defense; prestige and confidence in the U.S. scientific, technological, industrial, and military systems; and scientific observation and experimentation to add to our knowledge and understanding of the Earth, solar system, and universe.[2] They are still part of current policy discussions and influence the nation's civilian space policy priorities — both in terms of what actions NASA is authorized to undertake and the appropriations each activity within NASA receives.

NASA has active programs that address all four factors, but many believe that it is being asked to accomplish too much for the available resources. An understanding of how policy decisions made during the Sputnik era influence U.S. space policy today may be useful as Congress considers changing that policy. The response of Congress to the fundamental question, "Why go to space?," may influence NASA's programs, such as its earth-observing satellites, human exploration of the Moon and Mars, and robotic investigation of the solar system and wider universe as well as its policies on related activities, including spinoff technological development, science and mathematics education, international relations, and commercial space transportation.

This chapter describes Sputnik and its influence on today's U.S. civilian space policy, the actions other nations and commercial organizations are taking in space exploration, and why the nation invests in space exploration and the public's attitude toward it. The chapter concludes with a discussion of possible options for future U.S. civilian space policy priorities and the implication of those priorities.

SPUTNIK AND AMERICA'S "SPUTNIK MOMENT"

On October 4, 1957, the USSR launched Sputnik, the world's first artificial satellite. Sputnik (Russian for "traveling companion") was the size of a basketball and weighed 183 pounds (see figure 1). Sputnik's launch and orbit[3] still influences policy decisions 50 years later.

The USSR's ability to launch a satellite ahead of the United States led to a national concern that the United States was falling behind the USSR in its science and technology capabilities and thus might be vulnerable to a nuclear missile attack.[4] The resulting competition for scientific and technological superiority came to represent a competition between capitalism and communism.

Both the 85th Congress and President Eisenhower undertook an immediate set of policy actions in response to the launch of Sputnik. Congress established the Senate Special Committee on Space and Astronautics on February 6, 1958, and the House Select Committee on Science and Astronautics on March 5, 1958 — the first time since 1892 that both the House and Senate took action to create standing committees on an entirely new subject. Each committee was chaired by the Majority Leader. The Preparedness Investigating Subcommittee of the Senate Armed Services Committee was also active in analyzing the nation's satellite and missile programs.[5]

Multiple congressional hearings were held in the three months following Sputnik, and President Eisenhower addressed the nation to assure the public that the United States was scientifically strong and able to compete in space. Within 10 months after Sputnik's launch, the Eisenhower Administration and Congress took actions that

- established the National Aeronautics and Space Administration (NASA) through the National Aeronautics and Space Act (P.L. 85-568),[6]
- established the Defense Advanced Research Projects Agency (DARPA) within the Department of Defense through DOD Directive 5105.15 and National Security - Military Installations and Facilities (P.L. 85-325),[7]
- increased its appropriation for the National Science Foundation to $134 million, nearly $100 million higher than the previous year,[8] and
- reformed elementary, secondary, and postsecondary science and mathematics education (including gifted education) and provided incentives for American students to pursue science, technology, engineering, and mathematics postsecondary degrees via fellowships and loans through the National Defense Education Act (P.L. 85-864).[9]

Figure 2 provides a timeline of the some of the major policy events in the year following the Sputnik launch.

When people today speak of a "Sputnik moment," they often refer to a rapid national response that quickly mobilizes major policy change as opposed to a response of inaction or incremental policy change. The term is also used to question inaction — as in whether or not the nation is prepared to respond to a challenge without an initiating Sputnik moment.

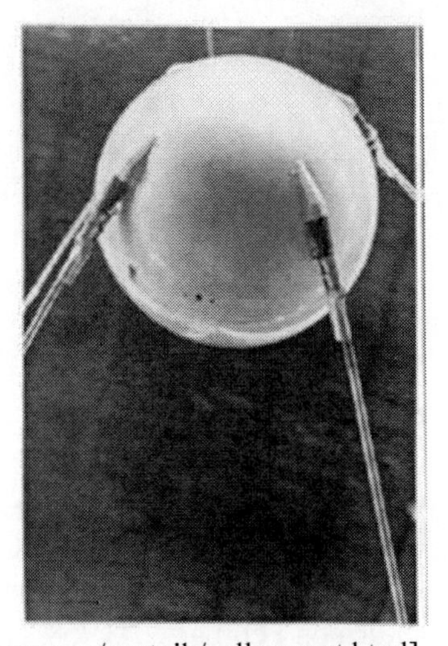

Source: NASA, at [http://history.nasa.gov/sputnik/gallerysput.html].

Figure 1. Sputnik

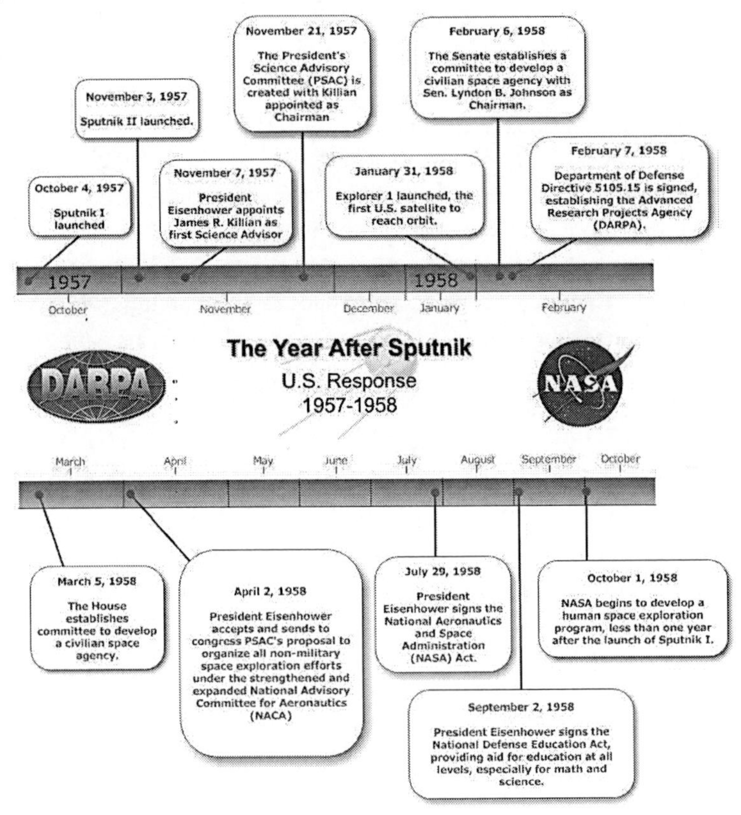

Source: Association of American Universities, at [http://www.aau.edu/education/ Sputnik_Timeline _2007-09-20.pdf].

Notes: DARPA was also established by Congress in P.L. 85-325.

Figure 2. Timeline of Select Policy Events in the Year Following the Sputnik Launch

Why Was Sputnik so Influential?

The Sputnik launch captured the public's attention at a time of heightened U.S. tension regarding the threat posed by the USSR and communism. Societal focus on civil defense, including "duck and cover" drills and the establishment of some personal bomb shelters, predisposed the nation towards identifying the potential threat posed by the Sputnik launch.[10] In this climate, many Americans became concerned that if the USSR could launch a satellite into space, it could also launch a nuclear missile capable of reaching the United States.[11]

The Sputnik launch was immediately viewed as a challenge to U.S. scientific and technological prowess. The Soviet Union launched both Sputnik and Sputnik 2 before the United States was able to attempt a satellite launch.[12] Additionally, the Soviet launch was of a far heavier satellite than the U.S. had planned.[13] The net result of the Sputnik launch was called a "Pearl Harbor for American Science" — a sign that the United States was falling behind the USSR in science and technology.[14] The ensuing competition in scientific and technological skills came to represent a competition to determine the political superiority of capitalism versus communism.

The Senate Majority Leader at the time, future President Lyndon B. Johnson, illustrated the concern of many Americans in his own observations of the night sky: "Now, somehow, in some new way, the sky seemed almost alien. I also remember the profound shock of realizing that it might be possible for another nation to achieve technological superiority over this great country of ours."[15]

Why Is Sputnik Important to Today's Policies?

The Sputnik launch prompted rapid development of new federal policies and programs. In particular, federal investment in NASA is still influenced by the Sputnik-era principles as illustrated in the Space Act, both in terms of what actions NASA is authorized to undertake and the extent to which each activity is funded.

In 2005, NASA was reauthorized for FY2007 and FY2008.[16] As Congress considers future reauthorization of NASA and the agency's 50th anniversary in 2008, the status of the nation's space policy, and the relative importance of the various objectives underlying this policy may become topics of debate.

The United States faces a far different world today than 50 years ago. No Sputnik moment, Cold War, or space race exists to help policymakers clarify the goals of the nation's civilian space program. The Hubble telescope, *Challenger* and *Columbia* space shuttle disasters, and Mars exploration rovers frame the experience of current generations, in contrast to the Sputnik launch and the U.S. Moon landings that form the experience of older generations.

What Are the Activities of Other Nations and the Commercial Sector in Space Exploration?

The United States faces a possible new set of competitors or collaborators in civilian space exploration. China, India, Japan, Russia, and Europe are taking an active role in space exploration as are commercial companies.[17] Japan's Moon orbiter was launched on September 14, 2007, and arrived at the Moon on October 4, 2007, the 50th anniversary of Sputnik. China's Moon orbiter was launched on October 24, 2007 and entered lunar orbit on November 5, 2007. India and the United States also plan to launch vehicles into the Moon's orbit by the end of 2008. The European Space Agency (ESA) is studying a Moon-lander mission, but its major focus is on Mars. The United Kingdom and Germany are independently proposing national Moon-orbiter projects, but their governments have not yet funded or approved these activities.[18]

China became the third country to send people into space, after the USSR[19] and the United States, completing two human spaceflight missions in 2003 and 2005.[20] Press reports indicate that China plans a manned Moon landing leading up to establishment of a Moon base after 2020,[21] though some have questioned these reports.[22] NASA Administrator Michael Griffin contends that China is likely to be the first country to land a human on the Moon since the last U.S. landing in 1972.[23] China has also indicated plans for a joint mission with Russia to send a satellite to orbit Mars in 2009.[24]

In the commercial sector, the X-Prize Foundation announced on September 13, 2007, the Google Lunar X Prize ($30 million) that invites private teams from around the world to build a robotic rover capable of landing on the Moon.[25] Virgin Galactic, currently based in California with a spaceport under construction in New Mexico, has plans for SpaceShipTwo, a six-passenger spaceliner with test flights scheduled for 2007 and suborbital passenger services to follow in 2008.[26] In Europe, EADS-Astrium is developing a four-person spacecraft to make suborbital trips with possibly the first commercial flight in 2012.[27] According to press reports, a number of venture capitalists are also planning to build spaceships or develop private space programs.[28]

If China is the first to land humans on the Moon and establish a Moon base in the 21st century or the European Space Agency is the first to land humans on Mars, will policymakers and the public view these activities as a loss in United States status and leadership? If so, what are the policy implications? Would such activities become this century's "Sputnik moment" that would spur further investment in U.S. space exploration activities? If not, how might this affect U.S. space policy priorities?

What is the Nation's Current Civilian Space Policy?

On August 31, 2006, President Bush announced a new U.S. National Space Policy.[29] The space policy defined the key objectives of defense and civilian space policy. The new space policy incorporated key elements of the Vision for Space Exploration ("Vision"), often referred to as the Moon/Mars program, previously announced on January 14, 2004.

In the Vision, President Bush announced new goals for NASA.[30] The President directed NASA to focus its efforts on returning humans to the Moon by 2020 and eventually sending them to Mars and "worlds beyond."[31] The President further directed NASA to fulfill commitments made to the 13 countries that are its partners in the International Space Station (ISS). In the 2005 NASA authorization act (P.L. 109-155), Congress directed NASA to establish a program to accomplish the goals outlined in the Vision, which are that the United States

- Implement a sustained and affordable human and robotic program to explore the solar system and beyond;
- Extend human presence across the solar system, starting with a human return to the Moon by the year 2020, in preparation for human exploration of Mars and other destinations;
- Develop the innovative technologies, knowledge, and infrastructures both to explore and to support decisions about the destinations for human exploration; and
- Promote international and commercial participation in exploration to further U.S. scientific, security, and economic interests.[32]

More specifically, the Vision includes plans, via a strategy based on "long-term affordability," to

- return the Space Shuttle safely to flight (which has been accomplished),
- complete the International Space Station (ISS) by 2010 but discontinue its use by 2017,
- phase out the Space Shuttle when the ISS is complete by 2010,
- send a robotic orbiter and lander to the Moon,
- send a human expedition to the Moon (sometime between 2015-2020),
- send a robotic mission to Mars in preparation for a future human expedition, and
- conduct robotic exploration across the solar system.[33]

NASA is developing a new spacecraft called Orion (formerly the Crew Exploration Vehicle) and a new launch vehicle for it called Ares I (formerly the Crew Launch Vehicle). An Earth-orbit capability is planned by 2014 (although NASA now considers early 2015 more likely) with the ability to take astronauts to and from the Moon following no later than 2020.

The Vision has broad implications for NASA, especially since almost all the funds to implement the initiative are expected to come from other NASA activities. Among the issues Congress is debating are the balance between NASA's exploration activities and its other programs, such as science and aeronautics research; the impact of the Vision on NASA's workforce needs; whether the space shuttle program might be ended in 2010; and if the United States might discontinue using the International Space Station.[34]

NASA states that its strategy is to "go as we can afford to pay," with the pace of the program set, in part, by the available funding.[35] Affording such a program is challenging, however, with a 2006 National Research Council report finding "NASA is being asked to accomplish too much with too little." The report recommended that "both the executive and the legislative branches of the federal government need to seriously examine the mismatch between the tasks assigned to NASA and the resources that the agency has been provided to accomplish them and should identify actions that will make the agency's portfolio of responsibilities sustainable."[36]

At a House Committee on Appropriations hearing, Dr. Michael Griffin, NASA's Administrator, acknowledged that funding is a challenge, stating

> Many people are amazed by the things that NASA accomplishes but don't realize that our budget is only 0.6 percent of the entire federal budget of the United States. As one of the most internationally recognized agencies in the government, many people assume that NASA's budget is much higher. In reality, we have to make tough choices in the allocation of scarce resources. We just cannot do everything that our many constituencies would like us to do. We need to set carefully considered priorities of time, energy and resources, and for this we're guided by the NASA Authorization Act of 2005, our annual appropriations, presidential policy, and the decadal surveys of the National Academy of Sciences. [37]

The House Committee on Appropriations also expressed concerns about these issues, stating NASA "has too many responsibilities and not enough resources to accomplish them all."[38] Similarly, the Senate Committee on Appropriations stated

> NASA's vision for space exploration maps out an aggressive role for the United States in manned space exploration. However, the potential costs are substantial and will likely be very difficult to maintain at the current estimated funding levels. The Committee is concerned that NASA will neglect areas that only tangentially benefit, or do not fit within, the exploration vision. The Committee believes that NASA must work diligently to balance existing programs and priorities with its plans for the future.[39]

Why Invest in Space Exploration?

The Appendix table compares The National Aeronautics and Space Act of 1958 as amended ("Space Act"),[40] the oldest and most recent Presidential commission reports (Killian[41] and Aldridge[42]), the U.S. National Space Policy[43] ("Space Policy"), and comments by members of Congress in the 110th Congress on the issue of why the United States might explore space. The analyses identify the following reasons why the United States might explore space:

- knowledge and understanding,
- discovery,
- economic growth — job creation and new markets,
- national prestige, and
- defense.

Some also include the following reasons:

- international relations, and

- education and workforce development.

Although there is broad agreement on the reasons for space exploration, there is a great deal of variation in the details. Among the chief differences in these documents are the degree to which

- discovery is the major reason for space exploration as opposed to meeting needs here on Earth;
- creation of jobs and new markets should be a major focus of NASA activities as opposed to a side effect;
- science and mathematics education and workforce development should be a goal of NASA in addition to other federal agencies; and
- relationships with other countries should be competitive or cooperative regarding space exploration.

According to an analysis conducted by the Space Foundation, the global space industry in 2005 generated $180 billion in revenues.[44] (See figure 3.) Comparing the Aldridge Commission themes, the Space Policy goals, and the Space Act objectives on the issue of the relationship of the space program to economic growth provides some insights. While the Aldridge committee has a much broader view of the industries related to space exploration, focusing on the potential role of space exploration in job generation and new market development, the Space Act and Space Policy focus on only one sector, the aeronautical and space vehicle industry.

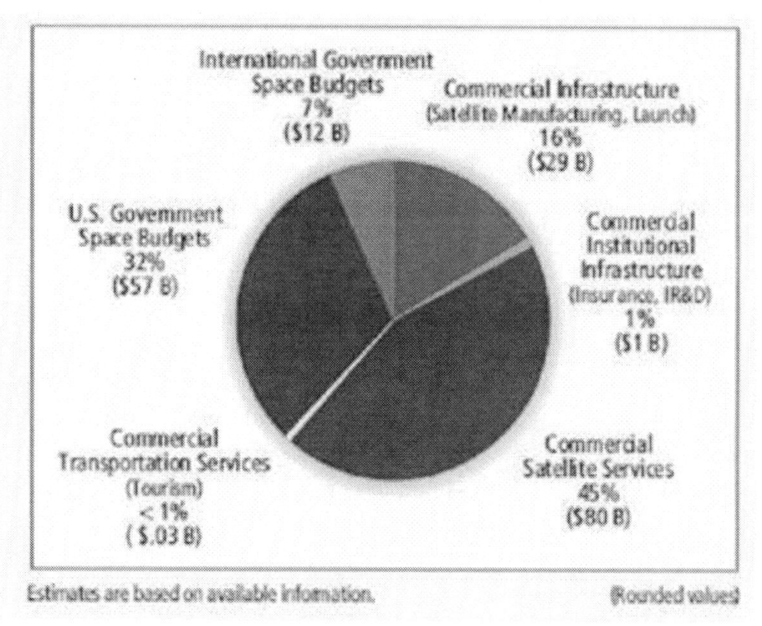

Source: Space Foundation, The Space Report: Guide to Global Space Activities, 2006, at [http://www.thespacereport.org/].

Figure 3. Global Space Economy Revenues, 2005

Table 1. NASA Centers

Center	Mission Area	Location
Ames Research Center	New Technology Research	Moffett Field, CA
Dryden Flight Research Center	Flight Research	Edwards, CA
Glenn Research Center	Aeropropulsion and Communications Technologies	Cleveland, OH
Goddard Space Flight Center	Earth, the Solar System, and Universe Observations	Greenbelt, MD
Jet Propulsion Laboratory (FFRDC)	Robotic Exploration of the Solar System	Pasadena, CA
Johnson Space Center	Human Space Exploration	Clear Lake, TX, near Houston, TX
Kennedy Space Center	Prepare and Launch Missions Around the Earth and Beyond	Cape Canaveral, FL
Langley Research Center	Aviation and Space Research	Hampton, VA
Marshall Space Flight Center	Space Transportation and Propulsion Technologies	Huntsville, AL
Stennis Space Center	Rocket Propulsion Testing and Remote Sensing Technology	Hancock County, MS, near Slidell, LA

Note: FFRDC is a federally funded research and development center.
Source: NASA, [http://education.nasa.gov/about/nasacenters/index.html]

The two Presidential commissions have two key differences. One is the first theme outlined in the Sputnik-era Killian Committee report: "the compelling urge of man to explore and discover." This is quite different from the recent Aldridge Commission report, which, although indicating exploration and discovery should be among NASA goals, states that "exploration and discovery will perhaps *not* be sufficient drivers to sustain what will be a long, and at times risky, journey." The implication is that, today, solely responding to the challenge of going to the Moon or Mars is not sufficient to energize public support for space exploration.

The second key difference is the focus of the Aldridge Commission on economic growth as a proposed space exploration theme. The Aldridge Commission identifies the ability of investments in civilian space programs to generate new jobs within current industries and spawn new markets. The contribution that federal space investments make to the nation's economy was not a key factor identified by the Killian Committee.

As a result of its focus on economic growth as a key theme of space exploration, the Aldridge Commission recommended that "NASA's relationship to the private sector, its organizational structure, business culture, and management processes —all largely inherited from the Apollo era — must be decisively transformed to implement the new, multi-decadal space exploration vision." Two of its specific recommendations were that NASA recognize and implement a far larger private industry presence in space operations, with the specific goal of allowing private industry to assume the primary role of providing services to NASA, and that NASA's centers be reconfigured as Federally Funded Research and Development Centers (FFRDCs) to enable innovation, work effectively with the private sector, and stimulate economic development.[45]

FFRDCs are not-for-profit organizations which are financed on a sole-source basis, exclusively or substantially by an agency of the federal government, and not subject to Office of Personnel Management regulations. They operate as private non-profit corporations, although they are subject to certain personnel and budgetary controls imposed by Congress and/or their sponsoring agency. Each FFRDC is administered by either an industrial firm, a university, or a nonprofit institution through a contract with the sponsoring federal agency. FFRDC personnel are not considered federal employees, but rather employees of the organization that manages and operates the center.[46]

NASA has not fully adopted the Aldridge Commission recommendations. NASA has 10 centers (see table 1).[47] One, the Jet Propulsion Laboratory (JPL), is already an FFRDC and is managed by the California Institute of Technology. In 2005 testimony before Congress discussing the involvement of private industry in NASA activities, NASA Administrator Griffin identified a need for continued federal capabilities. In particular, on the issue of private entrepreneurs who believe that they can launch vehicles and put payloads, including human beings, into space at a fraction of what it costs NASA, Griffin stated, "while I am enlisting the entrepreneurial community to step forward and help meet those requirements, we cannot stop work on the, admittedly less efficient, government systems.... [T]hat just doesn't work.[48] Similarly, when asked about whether or not NASA Centers would be converted to FFRDCs, Administrator Griffin stated, "I do not fundamentally see any gain to be achieved by having NASA convert federal centers to FFRDCs."[49]

WHAT IS THE PUBLIC'S ATTITUDE TOWARD SPACE EXPLORATION?

Some editorialists question whether investing in space exploration is relevant today.[50] Others question if NASA has the right priorities.[51] Would the public care if the country's investment in space exploration ended? Does the public believe it would be better to invest in social needs here on Earth rather than space exploration? Does the public support the current prioritization of the nation's space exploration activities?

According to poll data, Americans do not rank space exploration as a high priority for federal government spending. For example, in an April 10, 2007 Harris poll, respondents were given a list of twelve federal government programs and asked to pick two which should be cut "if spending had to be cut." Space programs led the list (51%), followed by welfare programs (28%), defense spending (28%), and farm subsidies (24%).[52] Space exploration was also near the bottom of a University of Chicago National Opinion Research Center survey reported in January 2007 that asked Americans about how they would prioritize federal spending.[53]

On the other hand, Americans are interested in space exploration. According to a Zogby poll reported in May 2007, most Americans (66%) are interested in space exploration with about half of those personally interested in commercial space travel and tourism. About half (49%) gave NASA a positive job approval rating. Many of those surveyed (45%) expressed concerns that the shuttle fleet is too old, too expensive, and too frail to fly humans into space safely. NASA's great achievements, according to poll respondents, are the Apollo Moon missions (34%) and the Hubble Space Telescope (18%).[54]

NASA's Office of Strategic Communication funded several analyses of the public's attitude toward space exploration based on focus groups[55] and a survey,[56] the results of which were presented in June 2007.[57] According to an analysis conducted for NASA, the focus group participants were ambivalent about going to the Moon and Mars and wanted to know why these missions were important. Reasons such as leadership, legacy, and public inspiration were found to be less persuasive, especially for future Moon exploration, than NASA-influenced technologies. Most participants agreed that partnership with other countries would be beneficial, but doubted whether it can be achieved realistically.

In addition, one of the analysis conducted for NASA found that most survey respondents rated NASA-influenced technologies[58] as somewhat or extremely relevant to them. Over 52% of participants said such technologies were a "very strong" reason to go to space. In contrast, the public's response to a mission to send humans to the Moon by the year 2020 was less strong with 15% of respondents very excited and 31% somewhat excited. Results for a mission to send humans to the Mars were similar to those for the Moon.

The public opinion analysis conducted for NASA found that there are generational differences in regard to NASA's proposed activities. For example, NASA's base support came from those who encompass "The Apollo Generation" (45-64 year olds), the majority (79%) of whom support NASA's new space exploration mission, particularly the return to the Moon. By contrast, the majority (64%) of those between 18-24 years of age are uninterested or neutral about a human Moon mission. Those between 25 and 44 years of age are approximately evenly split between those who are interested/excited and those who are either uninterested or neutral. Those over 65 were more likely to be neutral or disinterested in a Moon mission, with those over 75 years of age the least interested of all age groups.[59]

What are the Nation's Priorities for Civilian Space Exploration and Its Implications for Future Space Policy?

Current U.S. civilian space policy is based on a set of fundamental objectives in the Space Act, based on policy discussions that occurred following the launch of Sputnik over 50 years ago. Those objectives are still part of current policy discussions and influence the nation's civilian space policy priorities — both in terms of what actions NASA is authorized to undertake and the degree of appropriations each activity within NASA receives. NASA has active programs that address all its objectives, but many believe that it is being asked to accomplish too much for the available resources.

NASA was last reauthorized in 2005 for FY2007 and FY2008.[60] Thus, the reauthorization of NASA for FY2009 and beyond, along with NASA's upcoming 50th anniversary in Fall 2008, may provide an opportunity for Congress to rethink the nation's space policy. Some analysts have called for new 21st century space policy objectives and priorities to replace those developed 50 years ago.[61]

The goals of the nation's investment in space exploration may be a key factor in determining the focus of NASA's activities and the degree of funding appropriated for its programs. Congress and outside experts have concerns as to whether the United States can afford to implement President Bush's Vision for Space Exploration without adversely influencing NASA's other programs.[62] Congress may need to make challenging decisions to determine how to reap the most benefit from the nation's civilian space program investment. These decisions might answer questions such as

- What are the priorities among the many reasons for U.S. space exploration? For example, what might be the priority ranking among the previously identified reasons as to why the United States might explore space — knowledge and understanding, discovery, economic growth, national prestige, defense, international relations, and education and workforce development?
- What implications would this prioritization have for NASA's current and future budgets and the balance among its programs? For example, what is the proper balance between human and robotic space activities?
- What influence might the timing of other countries' space exploration activities have on U.S. policy? For example, what would be the impact of the United States, China, or another country, or a commercial organization, establishing the first Moon base or landing on Mars?

New objectives and priorities might help determine NASA's goals. This, in turn, might potentially help Congress determine the most appropriate balance of funding available among NASA's programs during its authorization and appropriation process.

For example, if Congress believes that national prestige should be the highest priority, they may choose to emphasize NASA's human exploration activities, such as establishing a Moon base and landing a human on Mars. If they consider scientific knowledge the highest priority, Congress may emphasize unmanned missions and other science-related activities as NASA's major goal. If international relations are a high priority, Congress might encourage other nations to become equal partners in actions related to the International Space Station. If spinoff effects, including the creation of new jobs and markets and its catalytic effect on math and science education, are Congress' priorities, then they may focus NASA's activities on technological development and linking to the needs of business and industry, and expanding its role in science and mathematics education.

Appendix Table. Possible U.S. Civilian Space Policy Objectives: Comparison of Selected Extracts from Historical and Current Space Policy Documents

Possible U.S. Civilian Space Policy Objectives	Space Act Objectives (July 1958, as amended)	Eisenhower Administration "Killian Committee" Factors (March 1958)	G.W. Bush Administration "Aldridge Commission" Themes (2004)	G.W. Bush Administration "Space Policy" Goals (2006)	Congressional Comments on Civilian Space Policy Objective-Related Issues in the 110th Congress
Knowledge and Understanding	"(1) The expansion of human knowledge *of the Earth* and of phenomena in the atmosphere and space."	"Fourth, space technology affords new opportunities for scientific observation and experiment which will add to our knowledge and understanding of the earth, the solar system, and the universe."	"Exploring the Moon, Mars, and beyond is a great journey worthy of a great nation. The impulse to explore the unknown is a human imperative, and a notable part of what animates us as a people. This endeavor presents an opportunity to inspire a new generation of American explorers, scientist, entrepreneurs, and innovators."	"Increase the benefits of civil exploration, scientific discovery, and environmental activities."	"From 2010, for another 3 to 5 years, we will have no access in space. We are going to rely on the kindness of allies to go back. We cannot lose time or ground. Our national security and our national honor depend upon it. Also, this would have a tremendous impact on the state of science, which goes to major efforts in terms of better understanding our planet Earth, where we do suspect intelligent life, and also the impact of climate changes."a (Senator Barbara Mikulski)
Discovery	"(2) The improvement of the usefulness, performance, speed, safety, and efficiency of aeronautical and space vehicles." "(3) The development and operation of vehicles capable of carrying instruments, equipment, supplies, and living organisms through space."	"The first of these factors is the compelling urge of man to explore and to discover, the thrust of curiosity that leads men to try to go where no one has gone before. Most of the surface of the Earth has now been explored and men now turn to the exploration of outer space as their next objective."	"Despite the spiritual, emotional, and intellectual appeal of a journey to space — exploration and discovery will perhaps *not* be sufficient drivers to sustain what will be a long, and at times risky, journey. We must also undertake this mission for pragmatic, but no less compelling reasons, which have everything to do with life here on Earth."	"Implement and sustain an innovative human and robotic exploration program with the objective of extending human presence across the solar system."	"...[H]ere we are on the 50th anniversary of Sputnik and it is another sputnik moment. When all of us in America were shocked that Russia had put up the first spaceflight, we were left to say: Why weren't we first? Today, 50 years later, we are looking at a 5-year gap from the end of the space shuttle before the crew-return vehicle will be on line to put American astronauts back in space. That is another Sputnik moment.

Appendix Table. (Continued)

Possible U.S. Civilian Space Policy Objectives	Space Act Objectives (July 1958, *as amended*)	Eisenhower Administration "Killian Committee" Factors (March 1958)	G.W. Bush Administration "Aldridge Commission" Themes (2004)	G.W. Bush Administration "Space Policy" Goals (2006)	Congressional Comments on Civilian Space Policy Objective-Related Issues in the 110th Congress
					Are we going to rely on Russia after 2010 to put American astronauts in space? I hope not. I hope America never loses its commitment to be the first in technology, in knowing what can be done, in exploring issues we haven't even thought about because we know how much that exploration has already done for our country."[b] (Senator Kay Bailey Hutchinson)
Economic Growth — Job Creation and New Markets	"(4) The establishment of long-range studies of the potential benefits to be gained from, the opportunities for, and the problems involved in the utilization of aeronautical and space activities for peaceful and scientific purposes."	Not discussed.	"Further space exploration will generate new jobs within current industries and will likely spawn entire new markets involving leading-edge manufacturing and flight support servicesAs one impressive labor leader testified to the Commission, 'every dollar spent on space is a dollar spent here on Earth.' This focus is good for jobs, good for the economy, and good for American families."	"Enable a dynamic, globally competitive domestic commercial space sector in order to promote innovation, strengthen U.S. leadership, and protect national, homeland, and economic security."	"NASA's space and Earth science basic research activities, along with microgravity research, are prime examples of research investments that cannot only advance our knowledge but benefit society [T]he investments also play a critical role in educating the next generation of scientists and engineers.
					Aeronautics R and D, we have discussed that at some great length, is another area where investments we make benefit our economy, our quality of life, and our national security Human space flight and

Appendix Table. (Continued)

Possible U.S. Civilian Space Policy Objectives	Space Act Objectives (July 1958, as amended)	Eisenhower Administration "Killian Committee" Factors (March 1958)	G.W. Bush Administration "Aldridge Commission" Themes (2004)	G.W. Bush Administration "Space Policy" Goals (2006)	Congressional Comments on Civilian Space Policy Objective-Related Issues in the 110th Congress
					exploration is another area that offers benefits ranging from the very important intangible inspiration it provides to the public, to the advanced technologies, and research results that can come from those initiatives"[c] (Congressman Mark Udall)
National Prestige	"(5) The preservation of the role of the United States as a leader in aeronautical and space science and technology and in the application thereof to the conduct of peaceful activities within and outside the atmosphere." "(9) The preservation of the United States' preeminent position in aeronautics and space through research and technology development related to associated manufacturing processes."	"Third, there is the factor of national prestige. To be strong and bold in space technology will enhance the prestige of the United States among the peoples of the world and create added confidence in our scientific, technological, industrial, and military strength."	"Exploring the Moon, Mars, and beyond is a great journey worthy of a great nation. The impulse to explore the unknown is a human imperative, and a notable part of what animates us as a people."	"Strengthen the nation's space leadership and ensure that space capabilities are available in time to further U.S. national security, homeland security, and foreign policy objectives."	"We are in a space race. While we are the current leader in space, there are many countries that want to take our place and are aggressively moving forward to do so...[O]ur Nation's space program benefits the lives of every American. The work that NASA does, from encouraging students into science and engi-neering careers, to innovative technology advances, improve our quality of life. The forward and innovative thinking at NASA helps to ensure our Nation has the ability to compete, and lead, in the global economy. We are committed to keeping our leadership role in space. In order to do so, we must make the right investments in space at the right times. That time is now."[d] (Senator Richard Shelby)

Appendix Table. (Continued)

Possible U.S. Civilian Space Policy Objectives	Space Act Objectives (July 1958, as amended)	Eisenhower Administration "Killian Committee" Factors (March 1958)	G.W. Bush Administration "Aldridge Commission" Themes (2004)	G.W. Bush Administration "Space Policy" Goals (2006)	Congressional Comments on Civilian Space Policy Objective-Related Issues in the 110th Congress
Defense	"(6) The making available to agencies directly concerned with national defense of discoveries that have military value or significance, and the furnishing by such agencies, to the civilian agency established to direct and control nonmilitary aeronautical and space activities, of information as to discoveries which have value or significance to that agency."	"Second, there is the defense objective for the development of space technology. We wish to be sure that space is not used to endanger our security. If space is to be used for military purposes, we must be prepared to use space to defend ourselves."	"Much of the United States' current military strength and economic security rests on our technological leadership. Our technological and industrial base must constantly be renewed. Therefore, the United States must continue to lead, especially in those industries that require, and therefore build, technology skills."	"Enable unhindered U.S. operations in and through space to defend our interests there." "Enable a robust science and technology base supporting national security, homeland security, and civil space activities."	"There is a strong, symbiotic relationship between space research and national security. For example, by using space-based navigation systems, we can guide a missile to within meters of its intended target. This not only allows our military to more effectively hit a target, it also saves civilian lives and limits collateral damage. The Chinese are gaining ground in technological areas. For example, China recently surpassed the U.S. as the world's largest exporter of information-technology products (and the U.S. has become a net importer of those products). The Chinese are now turning their attention to space technology — and they are determined to use it as a means of strengthening their military. We cannot allow other countries to acquire new weapons technologies while America does not keep up."[e] (Senator Bill Nelson)
International Relations	"(7) Cooperation by the United States with other nations and groups of nations in work done pursuant to this act and in	Not discussed.	"Although the era of Sputnik has given way to an age of international cooperation in space, it remains a competitive	"Encourage international cooperation with foreign nations and/or consortia on space activities that are of mutual benefit and that	"NASA has a successful histo-ry of international cooperation in science and involves non-U.S. partners on some two-thirds of its science missions, and also provides instruments,

Appendix Table. (Continued)

Possible U.S. Civilian Space Policy Objectives	Space Act Objectives (July 1958, as amended)	Eisenhower Administration "Killian Committee" Factors (March 1958)	G.W. Bush Administration "Aldridge Commission" Themes (2004)	G.W. Bush Administration "Space Policy" Goals (2006)	Congressional Comments on Civilian Space Policy Objective-Related Issues in the 110th Congress
	the peaceful application of the results thereof."		frontier...Other nations, against whom we compete for jobs in the global economy, are also intent on exploring space. If not us, someone else will lead in the exploration, utilization, and ultimately, the commercialization of space, as we sit idly by."	further the peaceful exploration and use of space, as well as to advance national security, homeland security, and foreign policy objectives."	science support, and other in-kind contributions to non-U.S.-led space and Earth science missions. Successful coopertive missions can increase the scientific content of a mission and build mutually beneficial relationships. At the same time, cooperation can lead to delays and added mission costs. Among the factors that have made international cooperative missions harder in recent years is ITAR" (International Traffic in Arms Regulations).[f] (House Committee on Science and Technology)
Education and Workforce Development	"(8) The most effective utilization of the scientific and engineering resources of the United States, with close cooperation among all interested agencies of the United States in order to avoid unnecessary duplication of effort, facilities, and equipment."	Not discussed.	"Long-term competitiveness requires a skilled workforce. The space exploration vision can be a catalyst for a much-needed renaissance in math and science education in the United States."	Not discussed.	"Madam President, we are observing the 50th anniversary of the launch of Sputnik, the first artificial satellite that was launched by humans...we suddenly became shocked at the fact that we were falling behind in math, in science and technology, and that, lo and behold, with the symbolic value of the Soviet Union — at that point our mortal enemy in the Cold War — having achieved that first. Now, there is a lesson in what I have just discussed about our history in space that would teach us not to repeat that now. What is that lesson?

Appendix Table. (Continued)

Possible U.S. Civilian Space Policy Objectives	Space Act Objectives (July 1958, as amended)	Eisenhower Administration "Killian Committee" Factors (March 1958)	G.W. Bush Administration "Aldridge Commission" Themes (2004)	G.W. Bush Administration "Space Policy" Goals (2006)	Congressional Comments on Civilian Space Policy Objective-Related Issues in the 110th Congress
					First of all, one of the great lessons of that era is the fact that we got excited about science and technology and mathematics and engineering and space flight. We produced a generation of exceptionally talented and educated young people who were told to go to their limit."[g] (Senator Bill Nelson)

Sources: "Space Act": P.L. 85-568, The National Aeronautics and Space Act, July 29, 1958. This analysis focuses on the objectives section. "Killian Committee": U.S. President (Dwight D. Eisenhower), President's Science Advisory Committee, *Introduction to Outer Space*, March 26, 1958. p. 2. Available at [http://www.hq.nasa.gov/office/pao/History/monograph10/doc6.pdf]. "Aldridge Commission": U.S. President (George W. Bush), President's Commission on Implementation of United States Space Exploration Policy, *A Journey to Inspire, Innovate, and Discovery*, June 2004, p. 11. Excerpts are from the section entitled "Why Go?". Available at [http://www.nasa.gov/pdf/60736main_M2M_report_small.pdf]. "Space Policy": U.S. President (G.W. Bush), U.S. National Space Policy, August 31, 2006, at [http://www.ostp.gov/html/US%20National%20Space%20Policy.pdf]. Excerpts are from section 3, "United States Space Policy Goals."

Notes: Excerpts are selected to reflect the general tone of text and are not necessarily the only language discussing these issues. The words in italics in the "Space Act" column show the changes made to the objectives since 1958.

[a] Senator Barbara Mikulski, "Departments of Commerce and Justice, and Science, and Related Agencies Appropriations Act, 2008," remarks in the Senate, Congressional Record, October 16, 2007, p. S12904.

[b] Senator Kay Bailey Hutchinson, "Departments of Commerce and Justice, and Science, and Related Agencies Appropriations Act, 2008," remarks in the Senate, Congressional Record, October 4, 2007, p. S12724.

[c] Congressman Mark Udall, U.S. Congress, House Committee on Science and Technology, Subcommittee on Space and Aeronautics, *NASA's Fiscal Year 2008 Budget Request*, opening statement, 110th Cong., 1st sess., March 15, 2007, H.Rept. 110-12 (Washington: GPO, 2007), p. 17.

[d] Senator Richard Shelby, "Departments of Commerce and Justice, and Science, and Related Agencies Appropriations Act, 2008," remarks in the Senate, Congressional Record, October 16, 2007, p. S12905.

[e]. Senator Bill Nelson, "Departments of Commerce and Justice, and Science, and Related Agencies Appropriations Act, 2008," remarks in the Senate, Congressional Record, October 4, 2007, p. S12726.

[f]. U.S. Congress, House Committee on Science and Technology, Subcommittee on Space and Aeronautics, *NASA's Space Science* Programs: Review of Fiscal Year 2008 Budget Request and Issues, 110th Cong., 1st sess., May 2, 2007, H.Rept. 110-24 (Washington: GPO, 2007), p. 6.

[g] Senator Bill Nelson, "Departments of Commerce and Justice, and Science, and Related Agencies Appropriations Act, 2008," remarks in the Senate, Congressional Record, October 4, 2007, p. S12726.

REFERENCES

[1] U.S. President (Dwight D. Eisenhower), President's Science Advisory Committee, *Introduction to Outer Space*, March 26, 1958. p. 1, at [http://www.hq.nasa.gov/office /pao/ History/monograph10/doc6.pdf].
[2] Ibid., 1-2.
[3] Sputnik 1 orbited the Earth every 96 minutes until it fell from orbit on January 4, 1958, three months after its launch. Roger D. Launius, "*Sputnik and the Origins of the Space Age,*" at [http://history.nasa.gov/sputnik/ sputorig.html].
[4] Dwight, D. Eisenhower Presidential Center, "Sputnik and the Space Race," at [http://www.eisenhower.utexas.edu/dl/Sputnik/Sputnikdocuments.html].
[5] U.S. Congress, House Committee on Science and Technology, *Toward the Endless Frontier: History of the Committee on Science and Technology, 1959-79,* prepared for the Committee by Ken Hechler, committee print, 96th Cong., 2nd sess., H.Prt. 35-120, (Washington: GPO, 1980), 1-28.
[6] P. L. 85-568, *The National Aeronautics and Space Act ("Space Act"),* July 29, 1958, at [http://www.nasa.gov/offices/ogc/about/space_act1.html].
[7] DARPA was originally called the Advanced Research Projects Agency (ARPA). It was established by DOD Directive 5105.15 on February 7, 1958, and by Congress in P. L. 85-325 on February 12, 1958. The name was changed from ARPA to DARPA by DoD Directive on March 23, 1972. DARPA was redesignated ARPA by President Bill Clinton in an Administration document on February 22, 1993. ARPA's name was changed back to DARPA by P.L. 104-106 on February 10, 1996. For more information about DARPA and its history, see DARPA, "Defense Advanced Research Project Agency: *Technology Transition*," January 1997 at [http://www.darpa.mil/body/pdf/transition.pdf].
[8] The appropriation for NSF continued to increase in future years. In 1968, it was almost $500 million. National Science Foundation, An Overview of the First 50 years, at [http://www.nsf.gov/about/history/overview-50.jsp].
[9] P. L. 85-864, *National Defense Education Act (NDEA)*, September 2, 1958.
[10]Dwight, D. Eisenhower Presidential Center, "*Sputnik and the Space Race,*" at [http://www.eisenhower.utexas.edu/dl/Sputnik/Sputnikdocuments.html].
[11]Council on Foreign Relations, Chronology of National Missile Defense Programs, June 1, 2002, at [http://www.cfr.org/publication/10443/].
[12]The Project Vanguard booster tests on December 6, 1957 (rose 3 feet, caught fire) and February 5, 1958 (rose 4 miles, exploded) were unsuccessful. A new effort, Project Explorer led by Wernher von Braun, was initiated. Explorer 1 was successful, after two aborted launches, on January 31, 1958. Roger D. Launius, "*Sputnik and the Origins of the Space Age*," at [http://history.nasa.gov/sputnik/sputorig.html].
[13]Sputnik 1 weighed 183 pounds. Sputnik 2 launched on November 3, 1957 weighed 1,120 pounds, carried a dog, and stayed in orbit for almost 200 days. The first satellite to be launched in the American Project Vanguard was planned to be 3.5 pounds. Roger D. Launius, "*Sputnik and the Origins of the Space Age*," at [http://history.nasa.gov/ sputnik/sputorig.html].

[14]Davis, James, C. *The Human Story: Our History, From the Stone Age to Today* (New York: Harper Collins, 2004). According to Davis, the statement was made by a Japanese newspaper shortly after the event. Others called it a "*technological Pearl Harbor*."

[15]Roger, D. Launius, "*Sputnik and the Origins of the Space Age,*" at [http://history.nasa.gov/sputnik/sputorig.html].

[16]NASA Authorization Act of 2005 (P.L. 109-155).

[17]Peter Spotts, "*Many Contestants in Latest 'Space Race' to the Moon,*" *Christian Science Monitor*, October 1, 2007.

[18]Frank Morring, Jr., Michael, A. Taverna and Neelam Mathew, "Nations Looking For a Piece of the Exploration Pie," *Aviation Week and Space Technology*, September 30, 2007, at [http://www.aviationweek.com/ aw/generic/story_channel.jsp?channel= space and id=news/ aw100107p2.xml and headline=Nations% 20Looking%20 For%20a%20 Piece% 20of%20the %20Exploration%20Pie].

[19]Since the fall of the USSR, Russia is the only former-USSR country to conduct space missions.

[20]See CRS Report RS21641, *China's Space Program: An Overview*, by Marcia S. Smith.

[21]*Agence France-Press*, "China Aims for Lunar Base after 2020," September 26, 2007, at [http://news.yahoo.com/s/afp/20070926/ wl_asia_afp/ spacechinamoon_070926154203]. Paul Maidmen, "China Shoots for the Moon," *Forbes*, August 14, 2007, at [http://www.forbes.com/ opinions/ 2007/08/14/notn-china-aerospace-oped-cx_pm_0814no tn.html].

[22]Dwayne Day, "*Exploding Moon Myths: Or Why There's No Race to Our Nearest Neighbor*," The Space Review, November 12, 2007, at [http://www.thespacereview.com/article/999/1].

[23]U.S. Congress, House Committee on Science and Technology, *NASA Fiscal Year 2008 Budget Request*, hearing, 110th Cong.,1st sess., March 15, 2007, H.Rept. 110-12, (Washington: GPO, 2007), p. 48. Michael D. Griffin, "The Role of Space Exploration in the Global Economy," speech, September 17, 2007, p. 41 at [http://www.nasa.gov/audience/formedia/speeches/ mg_ speech_ collection_ archive_ 1.html].

[24]"China and Russia Join Hands to Explore Mars," *People's Daily Online*, May 30, 2007, at [http://english.people.com.cn/200705/30/ eng20070530_ 379330.html].

[25]X-prize Foundation, *"Google Sponsors Lunar X PRIZE to Create a Space Race for a New Generation,"* press release, September 13, 2007, at [http://www.xprize.org/lunar/press-release/google-sponsors-lunar-x-prize-to-create-a-space-race-for-a-new-generation].

[26]Virgin Galactic, Overview, at [http://www.virgingalactic.com/]. NASA has signed a memorandum of understanding with Virgin Galactic to explore the potential for collaborations on the development of space suits, heat shields for spaceships, hybrid rocket motors, and hypersonic vehicles capable of traveling five or more times the speed of sound. See NASA, "NASA, Virgin Galactic to Explore Future Cooperation," press release, February 21, 2007, at [http://www.nasa.gov/home/hqnews/2007/feb/HQ_07049_Virgin_ Galactic.html].

[27]EADS-Astrium, *"Astrium Rockets into Space Tourism,"* press release, June 13, 2007, at [http://www.astrium.eads.net/press-center/press-releases/ astrium-rockets-into-space-tourism].

[28]Jeremy Quittner (ed.), "I Need My Space," *Business Week*, Winter 2007, at [http://www.businessweek.com/magazine/content/07_09/b4023413.htm].

[29]U.S. President (G.W. Bush), *U.S. National Space Policy*, August 31, 2006, at [http://www.ostp.gov/html/US%20National%20Space%20Policy.pdf]. It replaced the previous space policy that had been in place for 10 years.

[30]U.S. President (G.W. Bush), *President Bush Announces New Vision for Space Exploration Program, Fact Sheet: A Renewed Spirit of Discovery,* January 14, 2004, at [http://www.whitehouse.gov/ news/releases/ 2004/01/ 20040114-1.html].

[31]Twelve U.S. astronauts walked on the Moon between 1969 and 1972. No humans have visited Mars.

[32]U.S. President (G. W. Bush), "A Renewed Spirit of Discovery," document, January 14, 2004, at [http://www.whitehouse.gov/space/ renewed_ spirit.html].

[33]NASA, *Vision for Space Exploration*, February 2004, at [http://www.nasa. gov/mission n_pages/exploration/main/index.html].

[34]CRS Report RS22625, National Aeronautics and Space Administration: Overview, FY2008 Budget in Brief, and Key Issues for Congress, by Daniel Morgan and Carl E. Behrens; CRS Report RL33568, The International Space Station and the Space Shuttle, by Carl E. Behrens; and CRS Report RS21720, Space Exploration: Issues Concerning the "Vision for Space Exploration," by Marcia S. Smith.

[35]Testimony of Michael D. Griffin, Administrator, National Aeronautics and Space Administration before the Senate Committee on Commerce, *Science and Transportation Subcommittee on Space*, Aeronautics and Related Sciences, Budget Hearing, U.S. Senate, February 28, 2007, at [http://commerce.senate.gov/public/_files/ Testimony_Michael DGriffin_NASA_FY2008PostureStatementFINAL22707.pdf].

[36]National Research Council, Space Studies Board, *An Assessment of Balance in NASA's Science Program*, Washington, DC, 2006, p. 2 [http://www.nap. edu/ catalog.php? record_id=11644].

[37]Testimony of Michael, D. Griffin, Administrator, National Aeronautics and Space Administration before the U.S. House of Representatives, Committee on Appropriations, Subcommittee on Commerce, Justice, Science, *and Related Agencies*, FY2008 Budget Hearing, March 13, 2007, at [http://www.cq.com/display.do? dockey=/cqonline/prod/ data/docs/html/transcripts/congressional/110/congressionaltranscripts110-000002468892.ht ml@committees and metapub=CQ-CONGTRAN SCRIPTS].

[38]U.S. Congress, House Committee on Appropriations, report to accompany H.R. 3093, 110th Cong., 1st sess., July 25, 2007, H.Rept. 110-24, part 1 (Washington: GPO, 2007), p. 109.

[39]U.S. Congress, Senate Committee on Appropriations, report to accompany S. 1745, 110th Cong., 1st sess., June 29, 2007, S.Rept. 110-24, part 1 (Washington: GPO, 2007), p. 101.

[40]P.L. 85-568, The National Aeronautics and Space Act, July 29, 1958, at [http://www.nasa.gov/offices/ogc/about/space_act1.html]. Since 1958, the objectives have only had two modifications. The clause, "of the Earth and" was added to the first objective by the National Aeronautics and Space Administration Authorization Act, 1985, P.L. 98-361, § I 10(b), 98 Stat. 422, 426 (July 16, 1984). Objective (9) was added by the National Aeronautics and Space Administration Authorization Act, Fiscal Year 1989, P.L. 100-685, § 214, 102 Stat. 4083, 4093 (November 17, 1988). Objective (9) states the following: "The preservation of the United States' preeminent position in

aeronautics and space through research and technology development related to associated manufacturing processes."

[41]During the Sputnik era, President Eisenhower's Science Advisory Committee, chaired by George Killian, ("Killian Commission") responded to the fundamental question of why the United States might undertake a national space program in its report *Introduction to Outer Space*. (U.S. President (Dwight D. Eisenhower), President's Science Advisory Committee, *Introduction to Outer Space*, March 26, 1958. 2, at [http://www.hq.nasa.gov/office/pao/History/monograph10/doc6.pdf]). *The President's Science Advisory Committee is analogous to today's President's Council of Advisors on Science and Technology* (PCAST).

[42]U.S. President (George W. Bush), President's Commission on Implementation of United States Space Exploration Policy, *A Journey to Inspire, Innovate, and Discovery*, June 2004. Available at [http://www.nasa.gov/pdf/60736main_M2M_report_small.pdf]. The commission report is named for its chair, Edward C. "Pete" Aldridge, Jr., and called the "Aldridge Commission" report.

[43]U.S. President (G.W. Bush), *U.S. National Space Policy*, August 31, 2006, at [http://www.ostp.gov/html/US%20National%20Space%20Policy.pdf].

[44]Space Foundation, The Space Report: Guide to Global Space Activities, 2006, at [http://www.thespacereport.org/]. For more on the space economy, see Michael D. Griffin, Administrator, National Aeronautics and Space Administration, *The Space Economy*, NASA 50th Anniversary Lecture Series, September 17, 2007, at [http://www.nasa.gov/audience /formedia/speeches/mg_speech_collection_archive_1.html].

[45]U.S. President (George W. Bush), President's Commission on Implementation of United States Space Exploration Policy, *A Journey to Inspire, Innovate, and Discovery*, June 2004, p. 7, at [http://www.nasa.gov/pdf/60736main_M2M_report_small.pdf].

[46]See CRS Report RS21542, Department of Homeland Security: Issues Concerning the Establishment of Federally Funded Research and Development Centers (FFRDCs), by Michael E. Davey.

[47]For more information, see [http://education.nasa.gov/about/ nasacenters/ index.html].

[48]U.S. Congress, House Committee on Science, *Status of NASA's Programs*, H.Rept. 109-31, 109th Cong., 1st session, November 2, 2005, p. 53 (Washington, GPO, 2005).

[49]U.S. Congress, House Committee on Science, *The Future of NASA*, H.Rept. 109-19, 109th Cong., 1st session, June 28, 2005, p. 48 (Washington, GPO, 2005).

[50]See, for example, Anne Applebaum, "Mission to Nowhere," *Washington Post*, January 7, 2004, p. A21.

[51]See, for example, Gregg Easterbrook, "How NASA Screwed Up (And Four Ways to Fix It)," *Wired*, May 22, 2007, at [http://www.wired.com/science/space/magazine/ 15-06/ff_space_nasa]; *The Economist*, "Spacemen Are from Mars," September 27, 2007, at [http://www.economist.com/opinion/displaystory.cfm?story_id=9867224].

[52]Harris Interactive, "Closing the Budget Deficit: U.S. Adults Strongly Resist Raising Any Taxes Except "Sin Taxes" Or Cutting Major Programs," press release, April 10, 2007, at [http://www.harrisinteractive.com/ harris_poll/index.asp?PID=746]. The poll was of 2,223 adults surveyed online between March 6 and 14, 2007. This online survey is not based on a probability sample and therefore no theoretical sampling error can be calculated.

[53]University of Chicago, "Americans Want to Spend More on Education, Health," press release, January 10, 2007, at [http://www-news.uchicago.edu/releases/07/070110.gss.shtml]. The General Social Survey, supported by the National Science Foundation, has been conducted since 1973, and is based on face-to-face interviews of randomly selected people who represent a scientifically accurate cross section of Americans. For the 2006 survey, 2,992 people were interviewed and asked a wide variety of questions in addition to those related to spending priorities.

[54]Zogby, "Zogby Poll: Most Believe Humans Will Someday Colonize the Moon," press release, May 3, 2007, at [http://www.zogby.com/news/ ReadNews2.dbm?ID=1296]. The Zogby Interactive poll of 4,824 adults nationwide was conducted online from March 14-16, 2007, and carries a margin of error of ± 1.4 percentage points.

[55][55] The focus groups were professionally moderated by Dr. Stephen Everett of the Everett Group, Inc., in consultation with ViaNovo. The six focus groups were located in San Diego, Kansas City, and Philadelphia.

[56]The professionally conducted telephone survey was of 1,001 U.S. adults in February 2007. The margin of error was ± 3.2%. The survey was conducted by Dr. Mary Lynne Dittmar of Dittmar Associates, in consultation with ViaNovo.

[57]Robert Hopkins, "Strategic Communications Framework Implementation Plan," powerpoint presentation, NASA, Office of Strategic Communications, June 26, 2007, at [http://www.spaceref.com/news/ viewsr.html?pid=24646].

[58]An example of a NASA-influenced technology (commonly called "spinoff") mentioned in the survey that had significant results is a smoke alarm. According to NASA, in the 1970s NASA needed a smoke and fire detector with adjustable sensitivity for Skylab, America's first space station. Honeywell developed the device for NASA and then made it available commercially so that consumers could avoid "nuisance" alarms while cooking. Other devices in the survey were advanced breast cancer imaging, heart defibrillators, weather satellites, remote-controlled robots, global positioning system, cordless tools, satellite radio, and DirecTV. See [http://www.sti.nasa.gov/tto/] for more details on NASA's spinoffs. See a list of NASA's top 20 spinoffs in the last five years at [http://www.ipp.nasa.gov/spinoff_top_20a.pdf]. •

[59]Ibid., p. 9. Robert Hopkins, "Strategic Communications Framework,", powerpoint presentation, NASA, Office of Communications Planning, February 2007, at [http://images.spaceref.com/news/2007/ feb 07.stratcomm.pdf]. M. L. Dittmar, *The Market Study for Space Exploration*, (Houston, TX: Dittmar Associates, Inc., 2004), pp. 26-29 (age data) and pp. 8-11 (Executive Summary). M. L. Dittmar, "Engaging the 18-25 Generation: Educational Outreach, Interactive Technologies, and Space". Paper #2006-7303 in Proceedings of AIAA Space 2006, September 19-21, (San Jose, California. Washington, D.C.: AIAA, 2006). Paper and presentation available at [http://www.dittmar-associates.com/ Paper_ Downloads.htm].

[60]P.L. 109-155, NASA Authorization Act of 2005, December 30, 2005.

[61]See, for example, International Academy of Astronautics, "The Next Steps in Exploring Deep Space," June 9, 2004, at [http://iaaweb.org/iaa/ Studies/nextsteps.pdf]; Roger A. Pielke, " NASA Needs a New Vision," *San Francisco Chronicle*, October 7, 2007, p. E-5, at [http://www. sfgate.com/cgi-bin/article.cgi?file=/c/a/2007/10/07/edbjshsmh.dtl]; Do nald A. Beattie, "NASA's troubled future," The Space Report, September 4, 2007, at [http://www.thespacereview.com/article/945/1]; USA Today, "3 Alternate Visions for

NASA's Future," April 11, 2005, at [http://www.usatoday. com/tech/science/space/ 2005-04-11-space-alternate-visions_x.htm].

[62]See earlier discussion for Senate and House Committee on Appropriations report language; also Lennard Fisk, Chair, Space Studies Board, National Research Council and Thomas M. Donahue Collegiate Professor of Space Science, University of Michigan, *The President's Vision for Space Exploration: Perspectives from a Recent NRC Workshop on National Space Policy*, Testimony before the House Committee on Science, March 10, 2004, at [http://science.house.gov/Commdocs/hearings/ full04/mar10/ fisk.pdf].

INDEX

D

F

G

H

I

Q

R

S

T

U

V

W

X

Y